SOURCES
of
STATISTICS

SOURCES
of
STATISTICS

JOAN M HARVEY
FLA
Loughborough School of Librarianship

CLIVE BINGLEY   LONDON

MATH-STAT.

FIRST PUBLISHED 1969 BY CLIVE BINGLEY LTD
16 PEMBRIDGE ROAD LONDON W11
SET IN 10 ON 12 POINT LINOTYPE TIMES
AND PRINTED IN GREAT BRITAIN BY
THE CENTRAL PRESS (ABERDEEN) LTD
COPYRIGHT © JOAN M HARVEY 1969
85157 060 7

Contents

Introduction

The word 'statistics' appears first to have been used by Gottfried Achenwall in the middle of the eighteenth century. Then and for many years afterwards it was a general description of economic phenomena, not necessarily numerical facts systematically collected, which is the present definition of the word. Few statistics were, in fact, collected in Britain before 1946 and almost all that are available for that period were collected by government departments as part of the administration of the country. Since 1946 there has been a growing demand for statistics, and many more are being collected, analysed and published each year, not only by government departments but also by trade associations, professional bodies, local government authorities, individual firms and other organisations.

By no means all of the information collected is published. Sometimes this is because it is of a confidential nature, but often it is simply because it is unlikely to be of interest to enough people to warrant publication, or because it would be too costly to publish anyway. It is always worthwhile, therefore, to approach the organisation most likely to have collected the information, if no published information is available, or if what is published is not sufficiently detailed. Of course, some societies and associations rigidly restrict the supply of such information to their members, but others are more accommodating if the information can be shown to be of importance to the enquirer; others are only too happy to be able to help.

I have attempted to name and describe the main statistical publications of the United Kingdom, and have also included some of the more important United States publications and those of the various international organisations. In general, I have included only publications which contain a large proportion of statistics and have not, for instance, included the *Statesman's yearbook* or *Whitakers' almanack*, although these do contain many statistics taken from primary sources; nor have I included the *Statistical account of Scotland* which, despite its name, contains few statistical tables. Occasionally, publications which are mainly textual have been included because they are the only or the original source of particular information.

How reliable are statistics? Those mentioned in the following pages are all considered to be reliable sources but, when using statistics it is important to be aware of the reasons for which they were collected and analysed, what information was originally collected and what area it covered. Data collected for one purpose may be totally unsuitable for another. A description of the methods used, coverage, accuracy etc is usually incorporated into the preface or introduction to the volume or issue, or is published in a separate booklet. Footnotes can also be very important when placed at the foot of statistical tables and should not be overlooked; they may indicate that the figures are estimates or that some data has been excluded or some other equally important information.

No indication has been given as to whether government publications are parliamentary papers or in the non-parliamentary series. There is a tendency to remove titles of regular serials from the former to the latter series (*ie, Accounts relating to the trade and navigation of the United Kingdom* were issued as House of Commons papers, but the current *Overseas trade accounts of the United Kingdom* are not), and the division is somewhat artificial and unimportant, if not misleading, outside the realms of parliamentary procedure.

International publications have usually been given their English titles, and this is so even with the publications of the European Economic Community, even though the contents are not, with rare exceptions, in English.

There should be little difficulty in acquiring those publications which are generally available for purchase and in print. Some of the titles mentioned in this book, however, have a restricted circulation or are given to enquirers who first ask for the information. These can be difficult to track down and it is encouraging to note that the Board of Trade has decided that the various statistical publications concerning civil aviation shall no longer be restricted but be placed on sale to all through HM Stationery Office. Societies and associations providing statistical publications to members, often included in the price of the subscription to the organisation, usually have a non-members' price for the publications and there should be no difficulty in acquiring copies. I have deliberately tried to avoid mentioning publications which are without exception restricted to members of a particular association.

There should also be no difficulty in locating in libraries in the United Kingdom the more popular British and United Nations titles I have referred to. University and the larger public libraries are now taking more government publications but, unfortunately, there is a tendency to place an annual subscription for the parliamentary papers and to select only particular titles in the non-parliamentary series, which may be sound economically, but results in the library housing many publications which are never used and omitting to select other important titles. There seem to be very few collections of Northern Ireland government publications outside that country.

There are few sizeable collections of United States publications in the UK, and the lack of an agent here for their government publications makes acquisition more difficult. Apart from one or two of the larger public libraries, there is the collection of economic statistics at the Board of Trade's Statistics and Market Intelligence Library, and many libraries have copies of the *Statistical abstract*.

The United Nations deposits its documents in most of the larger libraries in the United Kingdom and there should be little difficulty in locating a good collection of United Nations publications, but the publications of the other international organisations are not so well represented, even though sales are handled by HM Stationery Office and it is not necessary to purchase direct.

So far as the other publications mentioned are concerned it is difficult to generalise, but the results of a survey of collections of economic statistics in United Kingdom libraries, designed jointly by the Library Association and Royal Statistical Society, should throw some light on the situation when the results are available in about two years' time. Certainly, there are too few good collections of statistical publications throughout the country, and if this book can in any way assist or persuade some libraries to acquire more representative collections it will have been well worth writing.

1*

General

The Permanent Consultative Committee on Official Statistics compiled an annual *Guide to current official statistics* from 1922 to 1938 (HMSO) but this was not revived after the war, although it had filled a very definite need. The Treasury produced a booklet *Government statistical services* in 1953, with a second edition in 1962, and some of the information it contained has been brought up to date by the Central Statistical Office's *List of principal statistical series available*: *I economic statistics, II financial statistics, III regional statistics* (HMSO, 1965) in the series 'Studies in official statistics'. The Interdepartmental Committee on Social and Economic Research have, as part of their work in examining the economic activities of particular government departments, produced six *Guides to official sources,* all of which are referred to in the appropriate chapters of this book. Other than these, there are the general sales lists of HM Stationery Office, none of which deal specifically with official statistical publications.

Many economic statistical publications are described in Ely Devons' *Introduction to British economic statistics* (CUP, 1956), F M M Lewes' *Statistics of the British economy* (Allen & Unwin, 1967) and R G D Allen's *Statistics for economists* (Hutchinson, 1952). *The sources and nature of the statistics of the United Kingdom,* edited by M G Kendall (Oliver & Boyd, 1952-57), containing articles first published in the *Journal* of the Royal Statistical Society, is still a useful guide in many subject fields, particularly for historical data. W Barker's *Local government statistics* (IMTA, 1965) is an excellent guide to statistics of local government finance and services.

In the United States, John L Andriot's *Guide to US government statistics* (Documents Index, third edition, 1961) is an annotated guide to all federal publications containing statistical data, and

Paul Wasserman's *Statistics sources* (Gale, second edition, 1965) is a classified listing of current statistical data designed to identify primary sources. *Statistical services of the United States government,* issued in 1959 by the Bureau of the Budget, describes the statistical system of the federal government and also the principal economic and social statistical series collected, whilst the same department's monthly *Statistical reporter* gives advance information on new statistical works, surveys and publications. Another Gale Research Company publication, *Statistical sources review,* reviews the statistical sources in periodicals, yearbooks, pamphlets and special reports by government and private agencies.

International bibliographies of statistics are scarce and, in the past, one has had to rely mainly on the general bibliographies such as the *United Nations documents index* and the general catalogues of publications issued by the United Nations and its agencies and other international organisations themselves. Professional journals occasionally include bibliographies, in particular *Revue de l'Institut International de Statistique.* As part of the Census Library Project the Library of Congress compiled some bibliographies, including *Statistical yearbooks of the world* (GPO, 1953) and *Statistical bulletins of the world* (GPO, 1954), and in 1955 the United Nations Statistical Office issued a *List of statistical series collected by international organisations,* but these are very much out of date now. In a more limited subject field, Joyce Ball has edited *Foreign Statistical documents* (Stanford UP, 1967), a bibliography of general international trade and agricultural statistics which, in part, claims to supplement W Gregory's *List of serial publications of foreign governments, 1815-1931* (Wilson, 1932), both of which are American compilations and exclude US publications. Joan M Harvey's *Statistics—Europe: sources for market research* (CBD Research Ltd, 1968) covers both east and western Europe, including the United Kingdom, and describes the statistical sources of each country which are useful for market research.

Some of the early *ad hoc* collections of statistical data can be useful for historical and sometimes unusual information. Mulhall's *Dictionary of statistics* (Routledge, fourth edition, 1899) includes information 'from the time of the Emperor Diocletian' to 1898 and includes a list of reference books although no authorities are given for the actual statistics quoted. This is supplemented by Webb's *New dictionary of statistics* (Routledge, 1911), arranged on the same general plan and covering the years 1899 to 1909, which does name authorities. Economic information is also given in compilations such as Macpherson's *Annals of commerce, manufactures, fisheries and navigation* (1805), Marshall's *Digest of all the accounts . . . relating to population, production and revenue* (1833), and Porter's *Progress of the nation* (1912).

General compilations of statistical data issued at regular intervals (annually, quarterly or monthly) contain a wealth of information in a convenient summarised form. The tables usually give a run of retrospective figures over several years, showing trends and enabling comparisons to be made. In general, more detailed statistics are published elsewhere, but occasionally one finds that the table in the general compilation is the only published information on the subject.

The *Statistical abstract,* containing ' potted ' figures on most topics, was issued annually by the Board of Trade from the first edition, covering the years 1840 to 1853, to the eighty third edition, covering the years 1924 to 1938, when it ceased publication because of the war. The series was revived after the war by the Central Statistical Office under the title *Annual abstract of statistics* (HMSO, 1946-). Currently, the publication contains sections devoted to statistical tables on area and climate, population and vital statistics, social conditions, education, labour, production, retail distribution and miscellaneous services, transport and communications, external trade, overseas finance, national income and expenditure, home finance, banking and prices. Many of these tables are summaries of statistics available in more detail in other publications, although some tables are not published elsewhere. Sources of information are given under each table and an index of sources, as well as a subject index, is included in each issue. Supplementing the *Annual abstract of statistics* is the *Monthly digest of statistics* (HMSO, 1946-), covering only the series of tables for which monthly or quarterly figures are available and therefore containing a smaller range of data although over a similar subject field. An annual supplement of definitions and explanatory notes is a valuable pamphlet which explains in some detail the definition, coverage and meaning of the monthly series included in both the *Monthly digest of statistics* and the *Annual abstract of statistics.*

A most useful volume of retrospective economic statistics is *Abstract of British historical statistics,* by B R Mitchell and Phyllis

13

Deane (CUP, 1962). Containing major time series data for the United Kingdom's economy for as long a period as possible (the earliest figure given is for 1199, and many tables begin with the eighteenth or nineteenth century and continue to 1938), each section is preceded by a short introduction and followed by a bibliography. Subjects covered include population and vital statistics, labour force, agriculture, coal, metals, textiles, transport, building, production, overseas trade, wages and standard of living, national income and expenditure, public finance, banking, insurance, and prices. Another useful publication, which covers a period for which regular statistical information is somewhat scarce, is *Statistical digest of the war* (HMSO, 1951), a general statistical volume in the series *History of the second world war* which brings together all the statistical information which illustrates the various individual volumes of the history. The data is mainly for the years 1939 to 1945 but 1936-38 averages are also given in some cases for comparison.

Over the past few years there has been a growing demand for regional statistics, both for market research and because of the Government's emphasis on regional development. In 1963 the National Economic Development Council referred to 'the paucity of systematic regional statistics' and the need for improvement, but there are considerable difficulties caused mainly by the variety of geographical boundaries used by organisations collecting statistics and by the need to co-ordinate the information collected so that it is compatible with other available information and with other regions. In 1965 the Central Statistical Office produced the first issue of its annual *Abstract of regional statistics* (HMSO, 1965-), which brings together statistics of the United Kingdom which are available on a regional basis, including population, employment, coal mining, fuel and power consumption, steel production, production generally, distribution, investment, foreign trade at principal ports, family expenditure surveys, etc.

The Scottish Statistical Office now compile twice a year a *Digest of Scottish statistics* (HMSO, 1953-) which includes tables on various aspects of industrial activity in Scotland, transport and communications, trade through Scottish ports, labour, population and vital statistics, social services, finance, family expenditure, distribution, and crime. The Welsh Office, also, compiles an annual *Digest of Welsh statistics* (HMSO, 1954-) which includes tables on population and vital statistics, social conditions, education, labour, production,

14

transport and communications, finance, and climate. The *Digest of statistics, Northern Ireland* (HMSO, 1954-), compiled by the Economic Section of the Cabinet Office at Belfast, is more comprehensive than the similar series for Scotland and Wales because more statistical information is collected for the territory separately. Issued twice a year, it contains data on population and vital statistics, labour, education, social conditions and services, fuel and power, production, building, agriculture, forestry, fishing, transport, postal services, imports and exports, finance, wages, prices, and retail sales. The *Ulster yearbook* (HMSO), now published annually, is also a useful source of statistics on all topics.

Regional economic statistics can also be found in the Institute of Marketing's *Basic economic planning data*, by E B Groves (the Institute, second edition, 1966) which includes data on population, insured employees, incomes, retail spending, car density, and rateable values, analysed by regions and counties as well as for the whole of the United Kingdom. It also summarises the economic characteristics of the twenty major cities of the United Kingdom having a population exceeding 200,000. Covering only one region of England is Gail G Wilson's *Social and economic statistics of North East England* (University of Durham, 1966) which deals with population, housing, rates and employment. Not primarily statistical but containing some useful statistical data are the reports of the regional planning councils now being published by the Central Office of Information for the Department of Economic Affairs, such as *The east midlands study* (HMSO, 1966), *A review of Yorkshire and Humberside* (HMSO, 1966), *Challenge of the changing north: a preliminary study* (HMSO, 1966), *The west midlands: patterns of growth* (HMSO, 1967), *A region with a future: a draft strategy for the south west* (HMSO, 1967), and *A strategy for the south east* (HMSO, 1967). *Scanning the provinces* (Northcliffe Newspapers Ltd, sixth edition, 1967) is a survey for advertisers and market executives of eight provincial areas.

The financial publications of local authorities often include some general statistics for the area, including population, employees, rates and rateable values, etc, but London and Birmingham also publish general statistical abstracts. The London County Council published a *Statistical abstract for London* for many years until 1951, and this was followed by *London statistics: new series,* the five volumes of which covered the years to 1960. The LCC has now

15

been replaced by the Greater London Council and the *1966 annual abstract of greater London statistics* (HMSO, or County Hall, London, 1968) has now been published, the information for the abstract having been collected and analysed by the GLC's Research and Intelligence Unit, formed to collect statistical materials of all kinds, to make statistical studies and analyses, and to forecast trends, among other activities. The first issue of the *City of Birmingham abstract of statistics* covered the years 1931 to 1949 and the tenth issue, covering several years to 1965 was published in 1967.

Economic trends (HMSO, 1953-), compiled monthly by the Central Statistical Office, contains tables and graphs of employment, output, consumption, prices, balance of payments, trade and finance, as well as special articles on economic topics of current interest. *New contributions to economic statistics* (HMSO, 1959-) contains some of the more important articles originally published in *Economic trends* and appears every two or three years. On an annual basis there is the 'blue book', *National income and expenditure*, which is dealt with in more detail in chapter eight, and the Treasury's *Economic report* (HMSO, 1962-), now published as a supplement to *Economic trends*. *Economic report*, which superseded the earlier *Economic survey*, contains a statistical appendix of mainly financial statistics, including balance of payments, expenditure on gross national product, imports and exports, gross domestic product, fixed capital formation, consumers' expenditure, price indices, and personal income, expenditure and saving. It, in fact, traces the course of the economy as a whole. The *Board of Trade journal* (HMSO, 1886-) still contains a number of statistical tables on various subjects, although many more are now published in the *Business monitor: production series,* and an index to the regular tables, indicating the issue and page in which they last appeared, is included in each issue of the periodical—usually on the last page.

Each issue of the quarterly journal of the National Institute for Economic and Social Research, *National Institute economic review,* (NIESR, 1959-), contains a section of regular tables of economic statistics, including one on productivity statistics, as well as economic articles which are often illustrated by statistical tables. A similar statistical service is provided by the quarterly *Bulletin* (1923-) of the London and Cambridge Economic Service. The publication has had a chequered career, having been published separately, as a supplement to the *Times review of industry*, and now in *The times* newspaper, the same information being available later in booklet form. This quarterly bulletin is now supplemented annually by a larger publication, *The British economy: key statistics, 1900-1966* (Times Newspapers Ltd, 1967) being the latest issue and including the best available figures for over 200 economic and social services which provide valuable background information on the long term development of the British economy.

From 1952 to 1966 the Colonial Office compiled a *Quarterly digest of colonial statistics* (HMSO), which gave data on population, foreign trade, production, national income and balance of payments, retail price indices, finance, education, etc for individual colonial territories. Some of the tables are being continued in the *Commonwealth Office yearbook* (HMSO, 1967-) but practically all countries, large and small, now issue their own quarterly or annual digests, bulletins or abstracts of statistics, compiled by the country's national statistical office.

The *Statistical abstract of the United States* (GPO, 1878-), compiled by the Bureau of the Census, is the standard summary of statistics on social, political and economic organisation of the country and is a veritable mine of information. It also contains a guide to other sources of statistical information, including census reports and statistical abstracts of individual states. (The abstract is now being published commercially, as well as officially, under the title *The US book of facts and statistical information*). Supplements to the official abstract include *Historical statistics of the United States: colonial times to 1957* and a later supplement which adds information to 1962; the *County and city data book*, which gives data for counties and cities over twenty-five thousand population; and the *Congressional district data book*. A new publication is the *Pocket data book, USA* (GPO, 1967-), which is to be a biennial compilation, less comprehensive than the abstract. Economic statistics, including employment, income, production, prices, etc, are published monthly in the US Joint Economic Committee's *Economic indicators* (GPO) and annually in a statistical appendix to the *Economic report of the President to the Congress* (GPO, 1947-). The revised edition of *Historical and descriptive supplement to economic indicators* (GPO, 1967) describes each of the series in the monthly publication and also gives annual data for years not shown in the monthly issues. A recent important

17

monograph which contains long historical series of statistics interspersed with the text is Emma S Woytinsky's *Profile of the US economy: a survey of growth and change* (Praeger, 1967), a comprehensive work covering every pertinent topic. Somewhat more specialised are *Long term economic growth 1860-1965: a statistical compendium* (GPO, 1966), a basic research document for economists, historians, investors, etc and the monthly *Business cycle developments* (GPO, 1961-), both issued by the Bureau of the Census, which brings together several hundred monthly and quarterly 'economic indicators' series for the analysis of short term economic trends and prospects, providing the basic data to facilitate such studies.

The most comprehensive international statistical publications are the United Nations' *Statistical yearbook* (UN, 1947-) and *Monthly bulletin of statistics* (UN, 1947-). The former title was first published by the League of Nations (1926-1942/44) and continued, with the 1947 issue, by the United Nations. It currently includes data for about 270 countries and territories, and covers a wide range of economic and social subjects, such as population, agriculture, manufacturing, construction, transport, trade, balance of payments, national income, education and culture. The latter title was also first published by the League of Nations but there was a gap in publication after 1944 until the United Nations recommenced publication in 1947. Less comprehensive than the yearbook, the bulletin covers fewer countries and subjects but is useful for its more up to date material. A supplement giving definitions and explanatory notes is published at intervals and sent to subscribers to the bulletin. UNESCO also issues a *Statistical yearbook* (UNESCO, 1963-) now, instead of the pocket size *Basic statistics,* which includes data on population, education, publishing, films, radio, and television. The Economist Intelligence Unit issues a series of sixty two *Quarterly economic reviews,* one for each country or area, which include statistics showing economic indicators and also foreign trade for certain items. Somewhat old now but still of value are the two monumental works by W S and E S Woytinsky, *World population and production* and *World commerce and government* (Twentieth Century Fund, 1953 and 1955).

Main economic indicators (OECD, 1965-) has replaced *Bulletin of general statistics* as the organisation's monthly statistical guide to recent economic developments and this is supplemented by an

18

occasional volume of historical statistics. Another separate quarterly supplement is devoted to industrial production statistics for the twenty-two member countries and this, too, has an occasional historical volume to supplement it. The Statistical Office of the European Communities publishes a *General statistical bulletin* (EEC, 1961-) and *Graphs and notes on the economic situation of the Community* (EEC, 1962-), both of which are monthly, and *Basic statistics of the Community* (EEC, 1960-) which is an annual publication. All these EEC publications provide general statistical information for each of the six Common Market countries, the latter title also giving comparative statistics for some other European countries, Canada, the USA and the USSR. *Statistical information,* published quarterly between 1953 and 1967, contained articles which usually included statistics. It is now replaced by *Statistical studies and surveys* (EEC, 1968-) which will contain accounts of surveys, articles on methodology, and qualified results not part of a regular series, initially in the social field.

CHAPTER TWO

Population, Vital Statistics and Migration

The first census of population of Great Britain was taken in 1801 and, except for the one due to be taken in 1941, censuses have been taken every ten years ever since. The earliest censuses were conducted by the Clerk of the House of Commons but from 1841 they have been the responsibility of the Registrar General. From 1861 a separate census for Scotland has been taken, which is the responsibility of the Registrar General for Scotland. Over the years the information collected has gradually increased and an account of the development of the census and of the information collected, analysed and published from 1801 to 1931 is given in the Inter-departmental Committee on Social and Economic Research's *Guides to official sources, no 2: census reports of Great Britain 1801-1931* (HMSO, 1951).

The latest full censuses for both England and Wales and for Scotland were taken in 1961 but a sample census was also taken in 1966. The main volumes of the *Census 1961: England and Wales* (HMSO, 1961-1966) contain age, marital conditions and general tables, industry tables, occupation tables, socio-economic group tables, housing tables, household composition tables, a report on the Welsh speaking population, scientific and technological qualifications, birthplace and nationality tables, Commonwealth immigrants in the conurbations, education tables, fertility tables, Greater London tables, migration tables, usual residence tables, and workplace tables. There are also separate reports for each county, a report on the Isle of Man, and a report on Jersey, Guernsey and adjacent islands. The volumes of the *Census 1961: Scotland* (HMSO, 1961-1966) include county reports and reports on Edinburgh and Glasgow, data on usual residence, housing and households, birthplace and nationality, occupations, Gaelic speaking population, internal migration, education, and fertility. *Great*

Britain, summary tables (HMSO, 1966) is a volume which summarises the information for all three countries, but it must be pointed out that the data collected for England and Wales and for Scotland is not always identical and it is often difficult, if not impossible, to arrive at a figure for Great Britain as a whole. Important accessories to the census reports are a two volume index of place names and a classification of occupations. As a result of the 1966 sample census a series of county reports and national reports covering the main census topics is being published for England and Wales and for Scotland, but whilst the information they contain is more up to date these reports are more limited in scope.

A separate census for Ireland was taken every ten years from 1801 to 1911. In 1926 a census was taken in both Northern and Southern Ireland, after which the General Register Office, Northern Ireland, was responsible for censuses taken in 1937, 1951 and 1961. The volumes of the 1961 *Census of population* (HMSO, Belfast) of Northern Ireland include separate reports on Belfast and the six counties, a fertility report, a general report and a topographical index.

The first census of population of the United States of America was taken in 1790 and the US Constitution requires the population of the country to be determined every ten years in order to apportion the representation of states in the House of Representatives. In 1909 the Bureau of the Census issued *A century of population growth: from the first census of the United States to the twelfth, 1790-1900* (GPO, reprinted 1966). The bureau's *Catalog of US census publications, 1790-1945* (GPO, 1950), which is kept up to date by quarterly and annual supplements, is useful for its listing of all US census publications. The latest census was taken in 1960 and the resulting reports contain data on the sex, age and marital status of the population, education, racial characteristics, occupations, etc, but many characteristics are collected in sample only. As the information collected is transferred to punched cards researchers can have particular items of information run off on payment of expenses. Special censuses are sometimes taken and the results published in series P28 of *Current population reports*.

The Department of Economic and Social Affairs of the United Nations, which was active in persuading most member countries to take censuses of population in or around 1960, has now issued *Principles and recommendations for the 1970 population censuses* and *Principles and recommendations for the 1970 housing censuses* (UN, 1967), and so it seems likely that most countries will now take a census of their population every decade.

DEMOGRAPHY

The General Register Office produces the *Registrar General's statistical review of England and Wales* (HMSO, 1921-) annually. Published in three parts, *Part I, tables, medical* is concerned with mortality and includes statistics of deaths by age and cause, still-births, and infectious diseases; *Part II, tables, population* gives statistics of births, marriages, fertility and also population estimates; *Part III, commentary* is mainly textual and comments on the figures of the other two volumes; *Decennial supplements,* issued after each census, give further detailed analyses. The *Annual report of the Registrar General for Scotland* (HMSO, 1855-) gives some-what similar information for Scotland. Updating some of the information in these publications are the *Registrar General's quarterly returns for England and Wales* (HMSO) and the *Quarterly returns of the Registrar General, Scotland* (HMSO), which cover births, deaths, marriages, infectious diseases and population esti-mates for their respective areas. Weekly, much less detailed returns are also issued. Estimates of population are also given in the *Registrar General's annual estimates of the population of England and Wales and of the local education authorities* (HMSO) and the *Annual estimates of the population of Scotland* (HMSO). A useful 100 year run of demographic statistics for the whole of Great Britain is included in the *Papers of the Royal Commission on population, 1950. Vol. II, reports and selected papers of the Statistics Committee* as ' Summary of demographic statistics for Great Britain ' (HMSO, 1950).

In Northern Ireland, the Registrar General for Northern Ireland publishes an annual report, quarterly returns and weekly returns containing similar information to the equivalent publications for England and Wales and for Scotland.

Vital statistics of the United States (GPO, 1937-), compiled annually by the National Center for Health Statistics, contains basic data on births, marriage, divorce and mortality for the country as a whole, for states and regions, and for Puerto Rico and the Virgin Islands. Demographic statistics are also published in series P20 and P25 of *Current population reports* (GPO), the former being concerned with population characteristics and the latter with population estimates.

The United Nations Statistical Office issues the *Demographic yearbook* (UN, 1948-) which is designed to supply basic statistical data for demographers, economists, sociologists and public health workers. Covering about 250 geographic areas of the world, the yearbook includes statistical data on all aspects of demography, the population of capital cities with over one hundred thousand inhabitants, and totals from the latest censuses. Some of the tables are brought up to date by the quarterly *Population and vital statistics report* (UN, 1949-). Vital statistics are also included in *World health statistics annual* (WHO, 1962-), which was previously the World Health Organization's *Annual epidemiological and vital statistics*.

In 1966 the Organization for Economic Cooperation and Development published *Demographic trends, 1965-1980, in Western Europe and North America* in two volumes, a general report and analyses by country, both of which include a number of statistical tables throughout the text. Some population statistics of the Common Market countries are given in *Basic statistics of the Community,* an annual handbook published by the Statistical Office of the European Communities, but they do not publish anything solely on population statistics.

MIGRATION

N H Carrier and J R Jeffrey's *External migration: a study of the available statistics, 1815-1950,* issued as no 6 in the General Register Office's ' Studies on medical and population subjects ' (HMSO, 1953) includes a large number of tables on overseas migration. Another study dealing with the United States as well as the United Kingdom, is Thomas Brinley's *Migration and economic growth: a study of Great Britain and the Atlantic economy* (CUP, 1954), which includes a chapter on statistical sources as well as many tables interspersed with the text. A special report on internal migration is included in M Newton and J R Jeffrey's *Internal migration* (HMSO, 1951).

Current annual statistics include the Home Office's *Statistics of persons acquiring citizenship of the United Kingdom and colonies* (HMSO, 1964-), *Statistics of foreigners entering and leaving the United Kingdom* (HMSO, 1939/51-), and *Control of immigration:*

statistics (HMSO, 1962-), the latter giving numbers admitted, embarked and refused admission.

The report of the Committee on Manpower Resources for Science and Technology, *The brain drain: report of the working group on migration* (HMSO, 1967) contains statistics on the estimated migration, both in and out, of scientists and technologists in the period 1961 to 1966. The *Annual abstract of statistics* includes statistics of external migration and the *Census of population reports* and also the *Registrar General's statistical review* have information on internal migration.

The *Annual report of the United States Immigration and Naturalization Service* (GPO, 1933-) contains statistical tables on immigration, aliens and citizens admitted and departed, and persons naturalised, and the same department issues a semi-annual *Report of passenger travel between United States and foreign countries.*

The United Nations' Department of Economic and Social Affairs produced an *Analytical bibliography of international migration statistics 1925-1950* (UN, 1955) which gives sources of international migration statistics for twenty-four countries, selected with a view to facilitating studies of emigration from Europe. Since 1962 statistics of international migration have been included in the United Nations' *Demographic yearbook* (UN, 1948-) but earlier figures can be obtained from *Sex and age of international migrants: statistics for 1918-1947* (UN, 1953) and *Economic characteristics of international migrants: statistics for selected countries, 1918-1954* (UN, 1958). A few tables showing migration statistics are usually included in the text of articles in the quarterly review of the Intergovernmental Committee for European Migration's *International migration,* which has been published since 1963 and which continued the Committee's *Migration* and the *REMP bulletin.*

CHAPTER THREE

Social Statistics

The earliest social statistics were published in the reports of social surveys and generally referred to particular towns or areas at particular dates. Works such as Charles Booth's *Life and labour of the people in London* (1892-1897), the London School of Economics' *The new survey of London life and labour* (1930-1935), and Seebohm Rowntree's surveys of York, *Poverty: a study of town life* (1901) and *Poverty and progress* (1941) are a valuable source of historical social information and contain many useful statistics. Since the war surveys have been made to help the planning of reconstruction and development of particular areas as well as on the social life of the people and these are often a source of statistical information.

SOCIAL SECURITY

The most recent guide issued by the Interdepartmental Committee on Social and Economic Research is *Guides to official sources, no 5. Social security statistics* (HMSO, 1961) which includes all statistical material collected by what was then the Ministry of Pensions and National Insurance and the National Assistance Board and is now the Ministry of Social Security, such as national insurance, family allowances, war pensions, and supplementary benefits.

The *Annual report of the Ministry of Social Security* (HMSO, 1966-) which replaced the *Report of the Ministry of Pensions and National Insurance* and the *Report of the National Assistance Board*, includes statistical tables on war pensions, contributions and benefits under the National Insurance Acts, family allowances, maternity benefits, death grants, retirement pensions, industrial injuries benefits, national assistance, legal aid, national insurance and family allowances abroad, and finance.

The *Annual report of the Ministry of Health* (HMSO, 1919/20-) includes some statistics throughout the text and also statistical appendices covering finance, general medical services, pharmaceutical services, general dental services, ophthalmic services, disciplinary action, local authority services, hospital and specialist services in England and Wales. Similar information for Scotland is given in *Health and welfare services in Scotland* (HMSO, 1962-), which is the annual report of the Scottish Home and Health Department, and *Scottish health statistics* (HMSO, 1961-), compiled by the Statistics Branch of the ministry, and for Northern Ireland in *Report on health and local government administration in Northern Ireland* (HMSO, Belfast), the annual report of the department. The annual *Local health services statistics* (IMTA & SCT) is concerned mainly with the cost of health services provided by county councils and county boroughs, whilst *National health service: hospital costing returns* (HMSO, 1950/51-) and *National health service, Scotland: analysis of running costs of hospitals* (HMSO, 1950/51-) both deal with hospital finances in particular.

Report on hospital in-patient enquiry (HMSO, 1960-) is a series, issued jointly by the General Register Office and the Ministry of Health, which has replaced the supplement to the *Registrar General's statistical review* on hospital in-patient statistics. The statistics are compiled from hospital case records of ten per cent of patients discharged from National Health Service hospitals, except psychiatric hospitals. *Scottish hospital inpatient statistics* (HMSO, 1963-) is compiled by the Scottish Home and Health Department. The Ministry of Health's *Report on the census of children and adolescents in non-psychiatric wards of National Health Service hospitals, June 1964 and March 1965* (HMSO, 1967) is an important *ad hoc* addition to hospital statistics.

Medical statistics are covered mainly by the General Register Office's publications mentioned in chapter two. Other than those, there is the Ministry of Power's annual *Digest of pneumoconiosis statistics* (HMSO, 1951-). A volume of *Medical statistics* is to be issued in the series *History of the second world war, United Kingdom medical service* but has not appeared at the time of writing.

The Interdepartmental Committee on Social and Economic Research's *Guides to official sources, no 3. Local government statistics* (HMSO, 1953) dealt mainly with current statistics collected and published by the Ministry of Housing and Local Government, although it did include a summary of some statistics collected by other government departments in England and Wales. *Local government statistics*, by W Barker (IMTA, 1965) also includes a chapter on housing and planning.

The *Annual report of the Ministry of Housing and Local Government* (HMSO, 1950-54-) always contained a number of statistical tables, but from 1965 it was decided to issue a *Handbook of statistics* (HMSO, 1965-) annually and a report as and when it is considered to be necessary. The handbook contains mainly housing statistics. Housing statistics for Scotland are included in the *Annual report of the Scottish Development Department* and for Northern Ireland in the *Report on health and local government administration in Northern Ireland* and in the *Ulster yearbook*. *Housing statistics: Great Britain* (HMSO, 1966-) is a new quarterly compiled jointly by the Ministry of Housing and Local Government, the Scottish Development Department, and the Welsh Office, giving data on new construction (including summary of progress, types of housing, areas and costs, housing densities, and contract size and type), improvement grants, slum clearance, housing loans etc. Similar information is given in *Housing return for England and Wales* (HMSO, 1946-), *Housing return for Scotland* (HMSO, 1946-), both issued quarterly, and the *Housing summary* (HMSO) issued jointly in intervening months and containing a brief statement of new houses built. The Institute of Municipal Treasurers and Accountants also publishes an annual return of *Housing statistics* (IMTA, 1949/50-) from information supplied by county boroughs, metropolitan boroughs, etc. *Housing return for Northern Ireland* (HMSO, Belfast) is a quarterly return of the Ministry of Health and Local Government, Northern Ireland, which shows progress on new housing, conversions and demolitions.

Statistics of decisions on planning applications (HMSO, 1962-) is compiled by the Ministry of Housing and Local Government and the Welsh Office from information supplied by all local planning authorities in England and Wales, and *County planning statistics* (SCT) is also issued annually.

27

Censuses of population include a considerable amount of information on housing, and there are also a number of tables in the *Annual abstract of statistics* and the *Monthly digest of statistics*. Further information on housing construction statistics is given in this book in the section on construction statistics in chapter six.

WELFARE

Welfare service statistics (IMTA & SCT) is compiled annually from information supplied by the various local government authorities in England and Wales. Welfare services in Scotland are dealt with in *Health and welfare services in Scotland* which has been referred to earlier in this chapter. The Home Office issues a *Report on the work of the Children's Department* (HMSO, 1961/63-) every three years and includes statistics of children in care, adoption, delinquency, remand homes, and approved schools. *Children's services statistics* (IMTA & SCT) is an annual mainly concerned with local government expenditure on children's services and the cost of maintenance in residential homes and nurseries. *Statistics relating to approved schools, remand homes and attendance centres in England and Wales* (HMSO, 1961-), issued annually by the Home Office, contains information on the numbers and categories of approved schools, admissions, numbers and ages of boys and girls in the schools, numbers sent to Borstals, success rate, length of stay, and nature of employment on release, etc.

CRIME

Here again there are separate publications for England and Wales, for Scotland and for Northern Ireland and comparisons can be difficult, mainly because the legal and judicial systems are not the same in all three countries. A useful article explaining the statistics and the system is Tom S Lodge's 'Criminal statistics' in *The sources and nature of the statistics of the United Kingdom*, edited by M G Kendall (Oliver & Boyd, 1952-57).

The Home Office's *Criminal statistics, England and Wales* (HMSO) and the Scottish Home and Health Department's *Criminal statistics, Scotland* (HMSO, 1925-) both contain data on the nature of the offences and the results of proceedings in the courts, appeals, probation orders, public prosecutions, extradition, legal aid, etc. The *Report on the administration of Home Office services (Northern Ireland)* (HMSO, Belfast) and the *Ulster yearbook* (HMSO) contain similar information for Northern Ireland. Supplementary unpublished information for England and Wales is available at the Home Office and can be obtained on request.

Prisons and borstals, 1963, statistical tables (HMSO, 1965) was followed by *Report on the work of the Prison Department* (HMSO, 1964-) and *Prisons in Scotland* (HMSO, 1927-), issued by the Scottish Home and Health Department, gives similar information for Scotland. Other Home Office statistical publications include *Offences of drunkenness* (HMSO, 1964-), which gives statistics of the number of offences of drunkenness proved in England and Wales, and *Offences relating to motor vehicles* (HMSO, 1950-), which is a return of the number of offences, prosecutions, court proceedings, etc, by police district.

The *Criminal Injuries Compensation Board report* (HMSO, 1964-65-), issued jointly by the Home Office and the Secretary of State for Scotland, includes statistical tables on applications resolved and compensation paid, outcome of awards and decisions made by single members, and analysis of no award cases.

POLICE

The *Annual report of HM Inspectors of Constabulary, England and Wales* (HMSO, 1880-), the *Report of HM Inspectors of Constabulary, for Scotland* (HMSO, 1858-), and the *Report of the Commissioners of Police of the Metropolis* (HMSO) all include some statistics on the establishment of the forces, offences, arrests, summonses, etc. *Police force statistics* (IMTA & SCT) is an annual compilation covering all police forces except the Metropolitan police and the City of London police, containing data on the strength of the establishments, houses provided, expenditure and income.

JUDICIARY

Civil judiciary statistics, England and Wales (HMSO, 1927-), compiled by the Lord Chancellor's Department, contains statistics relating to the Judicial Committee of the Privy Council, the House of Lords, the Supreme Court of Judicature, County Courts and other civil courts. Its Scottish equivalent, *Civil judiciary statistics (including licensing and bankruptcy)* (HMSO, 1935-), is compiled by the Scottish Home and Health Department and contains statistics relating to the House of Lords (Scottish appeals), the Court of Session, sheriff courts and other civil courts.

FIRE SERVICES

The *Report of HM Chief Inspector of Fire Services* (HMSO, 1948-) and the *Report of HM Inspector of Fire Services, Scotland* (HMSO, 1948/49-) both include statistics on the establishment of the fire brigades. Some information for Northern Ireland is given in the *Ulster yearbook* (HMSO). *United Kingdom fire services* (HMSO, 1960-) is an annual statistical analysis of reports of fires attended by fire brigades in the United Kingdom as a whole and is now compiled jointly by the Ministry of Technology and the Fire Office's Committee of the Joint Fire Research Organisation. Another return of statistics for fire authorities (county boroughs, counties, etc) in England and Wales is *Fire service statistics* (IMTA & SCT), which deals with the financial expenditure on fire brigades.

LICENCES

The Home Office issues two annual publications dealing with licensing statistics. *Licensing statistics* (HMSO, 1962-) has detailed statistics of the number of licensed premises (both on and off licences), proof gallons of spirits consumed per 100 population, gallons of beer consumed per head of population, and registered and licensed clubs. *Betting, gaming and lotteries act, 1963: permits and licences* (HMSO, 1963-) includes data on applications, cessations, etc of bookmakers' permits, betting agency permits, and betting office permits.

ELECTIONS

Review of elections (Institute of Electoral Research, 1954/58-) is published annually by the institute, each issue giving results of elections held since the previous issue.

United States publications in the field of social statistics include the Department of Health, Education and Welfare's monthly *Health, education and welfare indicators* (GPO, 1961-) and *Health, education and welfare trends* (GPO, 1961-), which is an annual. Both include charts and tables on US population, prosperity, health, radioactive fallout, school construction, pensions and insurance, unemployment, public assistance, and vocational rehabilitation. The Family

Services Bureau of that department issues a quarterly *Statistical report on social services* and the Children's Bureau an annual *Juvenile court statistics*. A *Census of housing* (GPO, 1940-) is taken as part of the *Census of population* and the 1960 report covers structural characteristics, building condition, occupancy, facilities, and financial aspects classified by states, counties, standard metropolitan statistical areas, rural areas, etc. Monthly and annual issues of *Housing statistics* are issued by the Housing and Urban Development Department and the Bureau of the Census' *Current construction reports* includes a series on housing starts. The bureau has also published *Housing construction statistics, 1889 to 1964* (GPO, 1966) which was designed as a historical supplement to the latter publication. The Federal Bureau of Investigation issues an annual *Uniform crime reports for the United States and its possessions* (GPO, 1930-) which is a statistical compilation of crime in the US, and the Bureau of Prisons issues periodic bulletins and reports on *National prisoner statistics*. *America votes: a handbook of contemporary American election statistics* (Macmillan, 1956-) is published biennially for the Governmental Affairs Institute.

The United Nations Statistical Office published in 1963 a *Compendium of social statistics* which, it was understood, was to be issued every three years but no second edition has appeared as yet. A valuable compilation, presenting basic national statistical indicators on major aspects of the social situation, it is divided into eight sections covering population and vital statistics, health conditions, food consumption and nutrition, housing, education and cultural activities, labour force and conditions of employment, social security, and income and expenditure. Explanatory notes describe each of the series and the difficulties which might make comparisons among countries hazardous. Social security statistics are also published in the *Yearbook of labour statistics* (ILO, 1951-), the United Nations *Statistical yearbook* (UN, 1947-) and *Statistical data,* issued by the Council of Europe. The World Health Organization's *World health statistics annual* (WHO, 1947-), the first fifteen volumes of which were titled *Annual epidemiological and vital statistics,* is now issued in three volumes concerning vital statistics and causes of death, infectious diseases, and health personnel and hospital establishments. It is updated in part by a monthly *Epidemiological and vital statistics report,* now *World health statistics report.*

The Statistical Office of the European Communities issues a series of *Social statistics*, each issue being devoted to a particular subject and usually concerned with labour problems, although no 4 of 1962 was on *Social security statistics, 1955-1960*.

CHAPTER FOUR

Education

There is now a considerable amount of statistical information available relating to education, which has been collected by government departments, local authorities, professional associations, etc, although until comparatively recently there was much less. A very useful article by Doris M Lee, 'The statistics of education', was published in the *Journal* of the Royal Statistical Society and later in *The sources and nature of the statistics of the United Kingdom,* edited by M G Kendall (Oliver & Boyd, 1952-57). A more recent guide to published statistics of education for the United Kingdom is contained in a chapter of W Barker's *Local government statistics* (IMTA, 1965).

The most comprehensive compilation is the annual *Statistics of education,* now issued by the Department of Education and Science (HMSO, 1961-). This publication has expanded rapidly in the past five or six years and the 1966 edition is to be in at least seven volumes. Volume one is devoted to statistics of schools, pupils, teachers, class size, immigrant children, etc, and covers maintained primary and secondary schools, nursery schools, direct grant and independent schools, special schools, and the school health service. The second volume contains data on GCE and CSE examinations, school leavers, GCE 'A' level results of students in further education, the flow of students with GCE qualifications, and the flow of school leavers. Volume three contains statistics of further education and adult education; volume four on the initial training of teachers and teachers in service; and volume five with finance (educational expenditure, salaries, etc), awards made by the Department of Education and Science and the local education authorities, school meals and milk, and educational building. All these volumes deal solely with England and Wales. Separate volumes on university statistics and on the deployment of teachers are planned.

Prior to 1961, summary statistics now included in the above publication were published in the annual report of the Ministry of Education (later the Department of Education and Science) (HMSO, 1928-) and some useful *ad hoc* statistical tables are still included in the text. The report for the year 1950, titled *Education 1900-1950* (HMSO, reprinted 1966), is a useful source of retrospective statistical information, including data on primary and secondary schools, further education (except university education), teachers, training of teachers, new buildings, health service, handicapped children, meals and milk, scholarships and other awards, and finance.

Selected statistics relating to local education authorities in England and Wales (list 71) (HMSO, 1956/57-64) contained data for each individual authority on the number of pupils, teachers, etc in primary and secondary schools, the school health service, school meals and milk, scholarships and other awards, and training colleges. Some of the information, including awards to students and entries to colleges of education, was subsequently incorporated in the 1965 edition of *Secondary education in each local education authority area (list 69)* (HMSO, 1956-), which also shows the flow of pupils resident in each area to the different types of secondary school. It is planned that the information given in *List 69* will eventually be published in the multi volume *Statistics of education*.

The Institute of Municipal Treasurers and Accountants and the Society of County Treasurers also publish an annual compilation, *Education statistics* (IMTA, 1950-). The statistical data is given by city, county and county borough of England and Wales and includes population, number of pupils at each type of school, average number of pupils in schools, number of schools maintained by local education authorities, pupil/teacher ratio, and finance. Data on finance covers the cost per pupil, cost of school dinners, cost of transport, and a comparative survey of the grant/revenue and expenditure of all local authorities.

The *Annual abstract of statistics* has summary tables, usually with a ten year run of figures, of the number of schools, pupils and teachers, training of teachers, further education, school meals and milk, school health service, finance, educational buildings, etc. Some regional statistics are given in the *Abstract of regional statistics,* and statistics for Wales only are shown in the *Digest of Welsh statistics.* Annual and quarterly statistics of educational building are also included in the *Monthly digest of statistics*. The *Census 1961. England and Wales.* *Education tables* gives data on a ten per cent sample basis relating to the terminal education age of persons resident in local areas and, for England and Wales and the regions and conurbations, classifications by age and occupation.

The main source of Scottish education statistics is *Education in Scotland*, the annual report of the Scottish Education Department (HMSO). This report deals comprehensively with Scottish education and the text is supplemented by a large number of statistical tables, covering expenditure on education, schools by type and size, pupils, classes, examinations, further education, school buildings and accommodation, teachers, and children in care. As for England and Wales, the *Annual abstract of statistics* has summary tables for Scotland and there are also tables in the *Digest of Scottish statistics*, whilst educational building statistics are in the *Monthly digest of statistics*. Volume nine of the *Census 1961: Scotland* gives data on a ten per cent sample basis relating to the terminal education age of the population.

Education in Northern Ireland, the annual report of the Ministry of Education, Northern Ireland (HMSO, Belfast), contains statistics of education in Northern Ireland, as does the *Digest of statistics, Northern Ireland* and the *Ulster yearbook.*

Education departments of local authorities collect statistics of schools, pupils and teachers; further education courses, students and teachers; courses and students in teacher training colleges; GCE and other examination results; scholarships; finance and educational building. Some authorities publish these statistics (*ie, London education statistics*) whilst others will make the information available to enquirers.

It is understood that, at the time of writing, consideration is being given to the possibility of issuing a joint publication by the Department of Education and Science, the Scottish Education Department and the Ministry of Education, Northern Ireland containing education statistics for the United Kingdom as a whole which, if it includes comparative data, would be a welcome addition to the publications already available.

The Robbins report, *Report of the Committee on Higher Education* (HMSO, 1963), contains many statistical tables relating to education at universities, colleges of advanced technology, training colleges and schools, both in the text of the report and in the separately published appendices. Statistics of university education have always been included in *Statistics of education* but with the 1966 edition it is anticipated that a whole volume will be devoted to this particular aspect of education, the information being prepared jointly by the Department of Education and Science and the University Grants Committee. The volume is likely to include data on the number of students entering each university, degrees and diplomas obtained, subjects taken, income and expenditure, etc previously published by the UGC in *Returns from universities and university colleges in receipt of Exchequer grant* (HMSO, 1947/48-). Also published by the UGC are the *Annual survey and review of university development* (HMSO, 1935/47-), which includes data on the number of students at each university and grants made to individual universities, and *First employment of university graduates* (HMSO, 1961/62-).

A new publication which is to appear annually is *Statistics of science and technology* (HMSO, 1967-), compiled by the Department of Education and Science and the Ministry of Technology, which brings together available statistics relevant to science and technology, including certain aspects of the training of scientists and technologists. Prior to this, the Council on Scientific Policy requested more comprehensive statistics on the scientific and technical qualifications of the population than had been available hitherto and, as a result, a question on the subject was included in the 1961 census questionnaire forms, the published report being *Census 1961. Great Britain. Scientific and technological qualifications* (HMSO, 1962).

Statistics of government expenditure on education are given in various publications, including the *Civil estimates, Civil appropriation accounts, Local government financial statistics, Local financial returns, Scotland,* and *Local authority financial returns, Northern Ireland,* all of which are dealt with in more detail in chapter eight. Mention must be made also of a new book by John Vaisey and John Sheehan, *Resources for education: an economic study of education in the United Kingdom, 1920-1965* (Allen & Unwin, 1968), which is about money spent on education in the last half century and contains an appendix of statistics.

The most important publication on education statistics of the United States is, perhaps, *Statistics of education in the United States,* compiled by the US Office of Education (GPO, 1958/59-).

36

An annual numbered series, each volume covers a particular area of educational statistics for the school year. This publication superseded the earlier *Biennial survey of education* (GPO, 1916/18-1956/58), but the statistical summary, which was included in the *Biennial survey,* is now published annually as the *Digest of educational statistics* (GPO, 1962-), containing current statistical information on schools, enrolment, teachers, graduates, educational attainment, finances, etc. Other publications include a *Research bulletin,* published four or five times a year by the National Education Association since 1923, which presents and interprets current statistical data drawn from the US Office of Education, the census figures, the research division of NEA itself, and reliable private sources.

Internationally, the *Statesman's yearbook* carries basic figures for education under each country entry, but more detailed information is available in three UNESCO publications. The *Statistical yearbook* (UNESCO, 1963-), which replaced *Basic facts and figures* (UNESCO, 1952-62), contains statistical data for over 200 countries on such items as number of schools, teachers and pupils, by sex, by level and type of education, by field of study, and by country of origin. Statistics of expenditure cover educational institutions, both public and private, but do not at present contain information on special, adult or other education not classified at levels. Each issue has statistics for a number of earlier years as well as for the latest year available, so developing a meaningful time series. The *International yearbook of education* (1933-), published jointly by UNESCO and the International Bureau of Education, includes a few statistical tables giving the latest available educational statistics of reporting countries (enrolments, numbers, and percentage of male and female pupils). The *World survey of education: handbook of educational organisation and statistics* (1955-) is produced triennially by UNESCO in four volumes and includes a few figures which are included to round out the picture for each country. The United Nations' *Statistical yearbook* also includes tables of statistics of education.

LIBRARIES

A critical review article on library statistics was published in the first issue of the Library Association's *Library and information bulletin* (LA, 1967-) and is well worth consulting regarding both United Kingdom and United States library statistics.

Currently, the official publication giving United Kingdom public library statistics is *Public library statistics* (IMTA), joint consultation on its preparation being held by the Institute of Municipal Treasurers and Accountants, the Society of County Treasurers, the Library Association and the Department of Education and Science. Information included in the publication is numbers of staff, service points, book stock, fines, income and expenditure, separate figures being given for each authority in the UK as well as national totals. Data on school library resources is also included. Prior to the 1963/64 issue, the data was confined to England and Wales, the Library Association publishing *Statistics of public (rate-supported) libraries in Great Britain and Northern Ireland.* The Library Association's statistics also included data on lending library issues and registered borrowers, information which is now no longer published. The annual reports of individual public library authorities may also include some statistical information.

Ad hoc publications including useful statistical information on public libraries are the Roberts report, *The structure of the public library service in England and Wales* (HMSO, 1959), which has a statistical appendix covering population, libraries, books, staff, expenditure and cooperation in 1957/58; *Standards of public library service in England and Wales* (HMSO, 1962), the report of a working party which contains sample statistics on expenditure, library materials, lending library use, staffing, etc, in 1960/61; *Inter library cooperation in England and Wales* (HMSO, 1962), the report of another working party which includes statistics of library cooperation in 1960/61; *Reference library stocks* and *Reference library staffs*, both issued by the Reference, Special and Information Section of the Library Association in 1960 and 1962 respectively. The *Annual report of the Council of the Scottish Library Association* includes statistics relating to the public libraries in Scotland, such as population served, staff, stock, issues and finance.

The University Grants Committee includes some statistics for university libraries in its *Returns from universities and university colleges* (HMSO, 1947/48-). Some statistics of stocks of books and periodicals in individual libraries of all kinds can be found in directories such as the Library Association's Reference, Special

Information Section's *Library resources* series, the Ministry of Defence *Guide to government departments and other libraries and information bureaux,* the *Aslib directory* and *Libraries, museums and art galleries yearbook* but, because of the nature of these publications, the data may not be current.

The Statistics Coordinating Project of the American Library Association recently published *Library statistics: a handbook of concepts, definitions and terminology* (ALA, 1966) and it is this volume which occasioned the review article in *Library and information bulletin* mentioned above. The US Department of Health, Education and Welfare issued some years ago *Statistics of public library systems serving populations of 100,000 or more, fiscal year 1960* and similar titles concerned with libraries of 50,000-99,999 and 35,000-49,999 populations, and library expenditure is included in the annual *Statistics of education in the United States.* As a result of a survey taken in 1966 the American Library Association have published *Library statistics of colleges and universities* (ALA, 1967) the information including data on staffing and material resources. Finally, *The Bowker annual of library and book trade information* (R R Bowker, 1956-) includes statistics on total library expenditure, library personnel and salaries in the US.

UNESCO's *Statistical yearbook,* referred to earlier in this chapter, contains statistical data for over 200 countries on the libraries and library service provided.

CHAPTER FIVE

Labour

Material collected by the Ministry of Labour was described in the Interdepartmental Committee on Social and Economic Research's *Guides to official sources, no 1. Labour statistics,* the latest revised edition being published in 1958 (HMSO). For a short time the guide was kept up to date by amendments published in the *Ministry of Labour gazette* but this is no longer done. However, the Statistics Department of the ministry does, from time to time, compile a list of the current statistical tables published by them.

The *Ministry of Labour gazette* (now *Employment and productivity gazette*) and *Statistics on incomes, prices, employment and production* are the main sources of UK labour statistics. The former (HMSO, 1893-), was first named the *Labour gazette* and later the *Board of Trade Labour gazette.* It is published monthly and includes a large number of statistical tables on manpower, employment, unemployment, training for employment, wage rates and hours of work, conditions of employment, industrial organisations, and the index of retail prices (often referred to as the cost of living index). An *Abstract of labour statistics* was published at intervals from 1894 until 1936, when it was suspended because of the war. After the war it was followed by *Tables relating to employment and unemployment* but this was only published from 1948 to 1950; since then there has been no annual compilation dealing solely with labour statistics. The Central Statistical Office's *Standard industrial classification* (HMSO), a consolidated edition of which was published in 1966, is used for labour statistics.

Statistics on incomes, prices, employment and production (HMSO, 1962-), which is prepared by the Ministry of Labour in collaboration with other government departments, principally the Board of Trade and the Central Statistical Office, is published quarterly. Its purpose is to make available in convenient form factual information which will assist those engaged in negotiation or arbitration to examine particular cases before them in relation to the wider implications of the decisions to be made. It includes data on wage rate, earnings, hours of work and other conditions of employment, manpower, prices, production, profits, and other relevant subjects. Much of the material is also published elsewhere but there are a few original tables.

Many labour statistics are also published in the *Annual abstract of statistics,* the *Abstract of regional statistics,* and the *Monthly digest of statistics.* Some labour statistics, such as data on occupations and place of work, are included in the censuses of population; the censuses of distribution and production have information about employment in the various trades and industries. Statistical publications on individual industries, such as the *Quarterly statistical review* of the cotton industry, the British Iron and Steel Federation and the British Iron and Steel Board's *Statistical yearbook* and *Monthly statistical bulletin,* and the *Annual report of the British Railways Board,* to name only a few, also include some labour statistics concerning their particular industries.

The University Grants Committee compiles an annual booklet, *First employment of university graduates* (HMSO, 1961/62-), although it does not, in fact, include information on graduates in all the disciplines. The Jones report, *The brain drain: report of the Working Group on Migration* (HMSO, 1967) contains statistics on estimated emigration and immigration of scientists and technologists between 1961 and 1966.

Time rates of wages and hours of work (HMSO) has been published at intervals and under various titles since 1893. After the 1929 issue there was a long gap until 1946, after which it has been published annually, except for the year 1953. The publication gives particulars of the minimum, or standard, rate of wages, the normal weekly hours, and conditions of employment in each industry, these having been fixed by voluntary collective agreements between organisations of employers and workpeople or by statutory orders under the Wages Council Acts and the Agricultural Wages Acts. The information is kept up to date by the monthly *Changes in rates of wages and hours of work.*

Statistics of stoppages of work due to industrial disputes are given monthly in the *Ministry of Labour gazette* and less detailed information is also given in the *Monthly digest of statistics* and

41

the *Annual abstract of statistics*. The number and membership of trade unions is published in the December issue of the *Ministry of Labour gazette* and also in the *Annual abstract of statistics*. Statistics of industrial accidents are published in the *Annual report of HM Chief Inspector of Factories* (HMSO, 1878/79-), and a *Guide to statistics collected by HM Factory Inspectorate* was published in 1960 (HMSO).

The Ministry of Labour, now renamed the Department of Employment and Productivity, has recently started a series of 'Manpower studies' designed to assist in the consideration of manpower policy as well as of economic planning more generally, and these studies often use primary statistical sources such as the census of population to show the structure and trends of certain occupations. Studies so far published have included one on the metal industries, one on the construction industry, and a more general one on occupational changes between 1951 and 1961 (HMSO). The National Economic Development Office is also issuing an occasional publication on manpower statistics, such as *Your manpower: a practical guide to the manpower statistics of the hotel and catering industry* (HMSO, 1967), which was produced by the University of Surrey mainly from the 1961 census and labour statistics.

The Statistics Department of the Ministry of Labour collects and analyses a considerable amount of statistical information, not all of which is of sufficiently general interest to warrant publication; it is often possible to obtain unpublished information from the department (Orphanage Road, Watford, Herts).

In 1953 the American Special Libraries Association published a revised edition of *A source list of selected labor statistics,* but this list has not been revised since that date. However, the Bureau of Labor Statistics of the US Department of Labor issues a six-monthly *Catalog of publications* which is not just a listing but an annotated subject index of all publications of the bureau issued during the six months covered by each issue. The bureau issues a large number of publications on labour statistics, some of which are mentioned below.

Monthly labor review (GPO, 1915-) is a publication of the bureau containing statistics of employment, work stoppages, payrolls, consumer prices and wholesale prices, as well as special articles on labour economics, working conditions and industrial relations. A biennial statistical supplement used to be issued, but this has been replaced by an occasional handbook, the latest of which is *Handbook of labor statistics* (GPO, 1967), which brings together historical series of statistics that bear on labour economics and labour institutions. Coverage is broad and includes labour force, employment and unemployment, hours of work, productivity, compensation, prices and living conditions, industrial injuries, and also some foreign labour statistics.

In 1966 two of the bureau's publications, *Employment and earnings* (GPO, 1954-66) and *Monthly report of the labor force* (GPO, 1959-66), combined to become *Employment and earnings and monthly report of the labor force* (GPO, 1966-), which presents current statistics on the labour force, employment, unemployment, hours, earnings and area data, as well as special articles. Publications carrying retrospective statistics are *Employment and earnings statistics for the United States, 1909-1967* (GPO, 1968) and *Employment and earnings statistics for states and areas, 1939-1966* (GPO, 1967), both of which were issued in the BLS bulletin series. Other important publications in this series are *Projections 1970: interindustry relationships, potential demand, employment* (GPO, 1967) and *Indexes of output per man hour: selected industries, 1939 and 1974-1965* (GPO, 1967). The *Manpower report of the President* (GPO, 1963-) also covers statistics of the labour force and is published annually.

The International Labour Office issues a *Yearbook of labour statistics* (ILO, 1941-) which presents a summary of the principal labour statistics for more than 170 countries or areas, including data on employment and unemployment, hours of work, labour productivity, wages, consumer prices, household budgets, industrial accidents, and industrial disputes. The office also issues a quarterly *Bulletin of labour statistics* (ILO, 1965-) containing statistics on employment and unemployment, wages, hours of work and consumer prices, which is updated by a supplement published each intervening month. The bulletin replaced a separate supplement which used to be issued with the *International labour review* (ILO, 1920-).

OECD has now published *Labour force statistics, 1956-1966* (OECD, 1968), which continues the earlier title *Manpower statistics* and is to appear regularly every two years in future. The publication aims to show the trend of the manpower and employment situation in member countries, data for each country including

43

population, changes in manpower, employment and unemployment, employment by activity and professional situation, salaries by industry, etc. The Statistical Office of the European Communities has recently published an annual volume of employment statistics in the series *Social statistics,* latterly as a supplementary volume. The same series also includes annual volumes on wages, on labour costs, and on occupational accidents in the iron and steel industries of the Common Market countries.

CHAPTER SIX

Production

AGRICULTURE: United Kingdom agricultural statistics have had a continuous history since 1865, when the first census of agriculture was taken. An article by Dennis K Britton and Kenneth E Hunt, 'Agriculture', which was reprinted in *The sources and nature of the statistics of the United Kingdom*, edited by M G Kendall (Oliver & Boyd, 1952-57) contains much useful information up to about 1950, both factual statistics and details of statistical sources. More recent is the Interdepartmental Committee on Social and Economic Research's *Guides to official sources, no 4. Agricultural and food statistics* (HMSO, 1958).

The main current publication is *Agricultural statistics: United Kingdom* (HMSO, 1939/44-), a joint annual compilation of the Ministry of Agriculture, Fisheries and Food, the Department of Agriculture and Fisheries for Scotland and the Ministry of Agriculture, Northern Ireland, giving the results of the annual agricultural censuses, acreage and production of crops, numbers of livestock, of agricultural workers, of agricultural holdings, and of certain descriptions of agricultural machinery, and price index numbers of agricultural products. Separate figures are given for England and Wales, for Scotland, and for Northern Ireland as well as for the United Kingdom as a whole. Separate publications are also issued for each of the three countries, *Agricultural statistics, England and Wales, Agricultural statistics, Scotland,* and the *Annual report of the Ministry of Agriculture, Northern Ireland. Output and utilisation of farm produce in the United Kingdom* (HMSO, 1946/47-), now issued separately at irregular intervals, was before 1946/47 part II of *Agricultural statistics: United Kingdom.* Summary figures of agricultural statistics are also published in the *Monthly digest of statistics.*

An *Annual review and determination of guarantees* (HMSO, 1947-) is compiled by the Ministry of Agriculture, Fisheries and Food and the Home Department for Scotland containing statistics on crops, livestock, workers, production, imports and costs. *Farm incomes in England and Wales* (HMSO, 1944/48-) is an annual survey of the financial characteristics of farming.

Dairy facts and figures (Federation of United Kingdom milk marketing boards, 1963-) is a useful compendium on dairy farming, milk supplies, utilisation, prices, transport, and advertising and sales promotion. Also published annually, with a six monthly supplement, is the *Statistical review* of the Fruit and Vegetable Preservation Research Association, which has data on the production, disposal, imports, etc of canned foods, quick frozen fruit and vegetables and fresh fruit and vegetables.

The latest US *Census of agriculture* (GPO) was taken in 1964. Preliminary reports have been issued and the final reports are being issued in three volumes—volume I, state and county statistics; volume II, general report; and volume III, special reports. *Agricultural statistics* (GPO, 1936-), issued annually by the US Department of Agriculture, is a reference book on agricultural production, supplies, consumption, facilities, costs and returns. The tables in each issue have runs for about ten years but the 1962 edition has earlier historical tables, from 1866 in some cases.

Publications of the Food and Agricultural Organization of the United Nations include *Production Yearbook* (FAO, 1947-), *Trade yearbook* (FAO, 1947-) and *Monthly bulletin of agricultural economics and statistics* (FAO, 1952-). *Production yearbook* contains data on land use, holdings, population, index numbers of agricultural production, crops, livestock, food supply, means of production, prices, and wages for each continent and for each reporting country. *Trade yearbook* deals with the foreign trade of reporting countries in agricultural products, and the monthly bulletin includes more up to date statistics on production, trade and prices of agricultural products as well as important articles on agricultural economics. *World crop statistics: area, production and yield* (FAO, 1966) is a historical volume covering much the same subject field as *Production yearbook* for the years 1948 to 1964. Finally, *Agricultural commodities projections for 1975 and 1985* (FAO, 1967) aims to define the scale and nature of the food problem that faces

46

the world and to assess long term prospects for world trade in major agricultural commodities.

World production, trade, consumption and prices of certain commodities, from the United Kingdom angle, are published in the annual reviews of the Commonwealth Secretariat, formerly the Commonwealth Economic Committee, *Grain crops, Dairy produce, Fruit, Meat,* and *Vegetable oils and oilseeds,* some of the information being updated by the various monthly or quarterly intelligence services they issue. In the past the publications were available from HMSO but now are available only from the secretariat.

The Statistical Office of the European Communities issues *Agricultural statistics* (EEC, 1961-), each of the eight or ten issues a year dealing with a particular aspect of agriculture, and *Agricultural prices,* issued monthly and annually and giving prices of the principal agricultural products in the Common Market countries and also in world markets.

Other organisations of all kinds publish helpful statistical information, the following being just a few. *Broomhall's corn trade year book: international grain trade statistics* (Northern Publishing Co Ltd) records the world's chief cereal crops, imports and exports and stocks; Frank Fehr & Co issue an *Annual review of oilseeds, oils, oilcakes, etc,* which includes data on production, imports and exports, prices and stocks of those products; the International Tea Committee issues an *Annual bulletin of statistics* and a *Monthly statistical summary,* both of which give data on the area under cultivation, production, exports, imports, consumption, stocks and auction prices of tea; C Czarnikow Ltd issue a weekly newsletter *Sugar review* and the International Sugar Council publishes *Sugar year book* and *Monthly statistical bulletin,* both of which deal with the production, import and consumption of centrifugal sugar; the Pan-American Coffee Bureau in New York has *Annual coffee statistics,* giving price movements, world trade and markets for coffee, and the Food and Agriculture Organization of the United Nations issues a quarterly *Cocoa statistics,* giving similar data for cocoa beans, cocoa butter, cocoa powder and paste, chocolate and chocolate products.

FISHERIES

The main publications are the annual *Sea fisheries statistical tables* (HMSO, 1888-), issued by the Ministry of Agriculture, Fisheries and Food and covering England and Wales, and *Scottish sea fisheries statistical tables* (HMSO, 1922-), issued by the Department of Agriculture and Fisheries for Scotland. Both include data on landings of fish, consumption level estimates, fishermen and fishing vessels, and imports and exports. *Fisheries of Scotland* (HMSO), the annual report of the department, also includes some statistics on fishing vessels, persons employed, lobster catch, escallop catch, mussels, shrimps, squids, herring, byproducts, prices and foreign landings of fish.

The *Annual report and accounts of the White Fish Authority* (HMSO, 1951/52-) includes statistical data on the quantity and volume of white fish landed, the value of shellfish landed, imports and exports of white fish, production and foreign trade of quick frozen fish, fish meal, the fishing fleet, the rate of catch, ownership of the trawlers, grant and loan schemes, the white fish subsidy, and the training of fishermen. The *Annual report of the Herring Industry Board* (HMSO, 1935-) also includes statistics of landings, imports, the composition and distribution of the fleet operating in East Anglia, grants and loans, earnings, and herring curing.

Fishery statistics of the United States (GPO, 1939-), compiled by the Bureau of Commercial Fisheries, gives comprehensive statistical data on the US fishing industry. Internationally, the Food and Agriculture Organization of the United Nations' *Yearbook of fishery statistics* (FAO, 1947-) gives information on catches and landings of fish and shellfish, and of production and trade in fish products, oils and fats, etc.

FORESTRY

The annual report of the Forestry Commissioners (HMSO, 1919/20-) includes some statistical tables on land use, forest land acquired, planting, thinning and felling, sales of timber, employment, and finance. There is also a table of statistics on this subject in the *Annual abstract of statistics* and publications dealing with timber, referred to later in this chapter, are also useful.

The Food and Agriculture Organization also publishes FAO *commodity review*, a general annual review of developments in the international commodity markets of the major agricultural commodities, including forest and fisheries products.

Index numbers of industrial production for both the United Kingdom and for Northern Ireland are published monthly in the *Board of Trade journal* and the *Monthly digest of statistics* and annually in the *Annual abstract of statistics*. A guide to this information is *Index of industrial production: method of compilation* (HMSO, 1959) which was issued by the Central Statistical Office and published as no 7 of their ' Studies in official statistics '.

The major sources of United Kingdom production statistics, which are not index numbers, are the *Reports of the census of production* (HMSO, 1907-). The census is described by Hector Leak in his article ' Censuses of production and distribution ' in *The sources and nature of the statistics of the United Kingdom,* edited by M G Kendall (Oliver & Boyd, 1952-57) and in the Interdepartmental Committee on Social and Economic Research's *Guides to official sources, no 6, census of production reports* (HMSO, 1961). Full censuses were taken for 1907, 1912 (which was not published because of the war), 1924, 1930, 1935, 1948, 1954, 1958 and 1963, the next one to be taken in 1969 on data for 1968. The reports of Import Duties Acts inquiries for 1933 and 1934 gave some information on the volume of production; a partial census was taken in 1946; and from 1949 simple sample censuses have been taken for each of the years between the full censuses. The two earliest censuses covered Great Britain and the whole of Ireland, and from then onwards the census reports have included data for Northern Ireland, except the report of the 1948 census, such data now being provided by the Ministry of Commerce, Northern Ireland. A full census of production covers manufacturing, mining, building and contracting, and public utilities, and the reports for each industry give an industry summary, an analysis by subdivisions of the industry, an analysis by size of enterprises, sales of principal products and of other products, employment, wages and salaries paid, the expenditure on plant and equipment. The report on the 1963 census is being issued in 133 parts including introductory notes, 128 parts each devoted to a particular industry, an index of products, and three parts containing summary tables. Where more detailed information is required than appears in the published reports, special analyses can often be undertaken by the Board of Trade's Statistics Division on payment of a fee to cover the cost of the work involved. However, such information must always be subject to strict observance of the rules preventing the disclosure of information relating to individual undertakings, as laid down in the Statistics of Trade Act, 1947, and this particularly restricts the amount of detail that can be given in an industry involving very few firms.

The census provides a framework on which up to date monthly and quarterly statistical enquiries can be based. Until 1962, such monthly and quarterly tables were published in the *Board of Trade journal* and/or the *Monthly digest of statistics,* some still appearing in these periodicals. In 1962, however, the Board of Trade started a new series of statistical publications, *Business monitor: production series* (HMSO, 1962-) in which there are about sixty separate titles published either monthly or quarterly, the data included varying according to the industry covered but usually including such information as orders in hand, new orders, production, deliveries, exports, raw materials used, and stocks. Current titles in the series are concerned with salt, pesticides and allied products, toilet preparations, paint and varnish, colours, soap and synthetic detergents, synthetic resins and plastics materials, polishes, engineers' tools, internal combustion engines, contractors' plant, refrigerating machinery, photographic equipment, scientific and industrial instruments and apparatus, watches and clocks, motor control gear, rotating electrical machinery, gramophone records, electric lamps, agricultural machinery, metal working machine tools, cars and commercial vehicles, wheeled tractors, trailers, motor cycles and pedal cycles, industrial fork lift and works trucks and tractors, locomotives, railway carriages and wagons, perambulators, tools and implements, cutlery, jewellery, domestic hollow ware, silk weaving and converting, linen, processed man-made continuous filament yarns, lace, hosiery and other knitted goods, carpets and rugs, narrow fabrics, clothing, gloves, footwear, timber, wood chipboard, bedding, pottery, electrical ware, floor coverings, rubber, brushes and brooms, paper and papermaking materials, toys and games, sports equipment, pens and pencils, boxes, stationery, safety razor sets and blades, domestic furniture, leathercloth, and abrasives. Subscriptions for the whole series or for single titles can be placed with HMSO, but single issues are only available from the Board of Trade libraries. Unfortunately, there is no detailed subject index to the series, nor any complete current list of titles, except in HMSO catalogues.

As well as being included in the census of production reports for the United Kingdom, the *Report of the census of production, Northern Ireland* (HMSO, Belfast, 1949-) is also published separately. The census is taken on a similar basis to that of the United Kingdom, the 1963 census comprising four volumes, a general report and summary tables; textiles and clothing; engineering, food, drink and tobacco, paper, printing and publishing; and timber and furniture, mineral products, other manufacturing trade, construction, gas, electricity and water.

In the United States there is an old established *Census of manufactures* (GPO, 1810-) which is now taken every five years by the Bureau of the Census, the results appearing first as preliminary reports, then as preprints, later in parts each referring to a single industry, and finally in a set of bound volumes. The final results of the 1963 census are published in three main volumes, summary and subject statistics, industry statistics, and area statistics, together with two more volumes on the location of manufacturing plants. The census is supplemented by an *Annual survey of manufactures* (GPO, 1943-), which is based on data from sample representative manufacturers and is published each year between censuses. A series of *Current industrial reports,* formerly titled *Facts for industry,* is roughly the US equivalent of the Board of Trade's *Business monitor: production series* referred to above. The bureau's *Guide to industrial statistics* (1964) is a comprehensive description of its programmes relating to industrial statistics.

Presenting internationally comparable data on the character and growth of the industrial sector of nearly one hundred countries is the United Nations' Statistical Office's *The growth of world industry* (UN, 1938/61-). It is to be published every three or four years and issued in two volumes, one containing national tables for each country on gross domestic product, industrial production indices, characteristics, etc, and the other containing international analyses and tables on factors and patterns of growth in industry and the structure of economic activity.

The Organization for Economic Cooperation and Development's *Industrial statistics, 1900-1962* (OECD, 1964) was planned as a reference work containing the more important industrial production data for member countries. It is updated by the industrial statistics published in *Main economic indicators,* referred to in more detail in chapter one. The Statistical Office of the European Communities issues a quarterly and annual *Industrial statistics* (EEC, 1960-), which has index numbers of industrial production in Common Market and some other countries and production statistics for a wide range of commodities produced in Common Market countries.

FUEL AND POWER

The Ministry of Power's annual *Statistical digest* (HMSO, 1938/43-) covers production, distribution, capital expenditure and employment in the various fuel industries, coal, coke, manufactured fuel, electricity, gas, benzole, coal tar, and petroleum. Information on natural gas at present includes only data on licences issued. Information on sources of earlier statistics is usually given in the introduction to each annual volume. A *Weekly statistical statement,* giving provincial figures on coal, gas and electricity is available from the Ministry of Power, and monthly statistics or weekly averages for those industries and for petroleum are given in the *Monthly digest of statistics.* The *National Coal Board report and accounts* (HMSO, 1946-) now consists of an annual report and a second annual volume of accounts and statistical tables, the tables including data on output, consumption, stocks, and exports of coal, and manpower and average earnings in the industry. A quarterly statistical statement of the cost of production, proceeds and profits of collieries, published first by the Ministry of Power and later by the board, was discontinued a few years ago. The Central Electricity Generating Board now issues a CEGB *statistical yearbook* (CEGB, 1964-) which supplements the information given in the board's annual reports, a statistical appendix having previously been incorporated in the reports. Some statistical information is included in the annual reports of the various regional electricity boards, and also in the annual reports of the Gas Council and of the regional gas boards, all of which are published by HMSO.

The Petroleum Information Bureau includes two statistical annuals among its publications. *UK petroleum industry statistics* gives information on the consumption and production of refined products in the United Kingdom, and *World oil statistics* gives information on production, refining capacity, consumption, tanker tonnage, etc of oil by countries. They also issue a number of area booklets which contain tables of production and refinery capacity statistics relative to the countries in each area, and will also try to answer *ad hoc* enquiries of a non technical nature about petroleum.

Two of the many important American statistical publications on fuel and power are the *Semi annual electric power survey,* the results of which are published by the Edison Electric Institute and include data on the electric power supply and the manufacture of heavy electric power equipment in the us, and *Petroleum facts and figures,* issued biennially by the American Petroleum Institute and designed to bring together in a single volume the most complete and comprehensive record available of the petroleum industry's operations in the us and in individual states.

The United Nations' Economic Commission for Europe issues a *Quarterly bulletin of coal statistics for Europe* (UN, 1952-), which contains data on production and stocks of various types of coal and on imports and exports for European countries and the USA, an *Annual bulletin of electric energy statistics for Europe* (UN, 1958-), and a *Half yearly bulletin of electric energy statistics for Europe* (UN), which have data for European countries and USA on plant consumption, thermal capacity, production and consumption of electricity. The last title was until recently a quarterly publication. The Organisation for Economic Cooperation and Development publishes *Basic statistics of energy* (OECD, 1952/63-) covering data on the OECD area as a whole and for the individual member countries, and *Oil statistics* (OECD, 1961-), which is published annually and is up dated by *Provisional oil statistics by quarters* (OECD, 1964-), both dealing with the production, supply and disposal, imports and exports, and consumption of oil by end users for each member country. The Statistical Office of the European Communities issues a quarterly and an annual volume of *Energy statistics* (EEC, 1953-) covering coal, coke, lignite, gas, petroleum and petroleum products, and electrical energy, and showing the principal indicators of the energy situation and overall energy balance sheets as well as statistics by source of energy, which was originally issued by the European Coal and Steel Community.

MINERALS AND METALS

A comprehensive annual volume on the mineral industry of the world, including the United Kingdom, is compiled by the Mineral Resources Division of the Overseas Geological Surveys. *Statistical*

53

summary of the mineral industry: world production, exports and imports (HMSO, 1913/20-) contains data on quantities produced, imported and exported of particular minerals so far as the information is available. The minerals concerned are abrasives, aluminium, antimony, arsenic, asbestos, barium minerals, bentonite, beryl, bismuth, borates, bromine, cadmium, cement, china clay, chrome ore and chromium, coal, coke and by-products, cobalt, copper, diamonds, diatomaceous earth, feldspar, fluorspar, Fuller's earth, gold, graphite, gypsum, iodine, iron and steel, lead, lithium minerals, magnesite and dolomite, manganese, mercury, mica, molybdenum, nickel, nitrogen compounds, petroleum and allied products, phosphates, platinum, potash minerals, pyrites, rare earth and therium minerals, salt, selenium, sillimanite, silver, strontium minerals, sulphur, talc, tantalum and niobium minerals, tin, titanium minerals, tungsten, uranium minerals, vanadium, vermiculite, zinc, zirconium minerals, and other minerals and metals.

Statistics of the mineral industry are well documented in the United States. The latest *Census of mineral industries* (GPO, 1840-) is the report of a census conducted jointly with the 1963 census of manufactures, and the US Department of the Interior, Bureau of Mines, issues a four volume *Minerals yearbook* (GPO, 1882-) covering production, consumption, stocks, transport, labour and productivity, prices and costs, income and investment, which includes many statistical tables. The volumes are concerned with metals and minerals (except fuels), mineral fuels, regional reports within the USA, and a volume concerning other countries of the world. Updating this is a series of monthly or quarterly *Mineral industry surveys,* each on a particular mineral.

The International Tin Council issues a *Statistical yearbook* (the Council, 1959-) covering production, consumption, imports and exports of tin, tinplate and canning. Sometimes a supplement is issued instead of a full annual volume.

A useful quick reference book for iron and steel and non ferrous metals is the annual *Quin's metal handbook* (Metal Bulletin Ltd, 1913). The Iron and Steel Board and the British Iron and Steel Federation issue *Iron and steel: annual statistics for the United Kingdom* (BISF, 1959-) and *Iron and steel: monthly statistics* (BISF, 1956-), both of which contain detailed statistics of all aspects of the industry in the United Kingdom, the monthly also including data on overseas countries. The federation also issues an annual two volume looseleaf *Statistical handbook* (BISF, 1959-) containing a section for every iron and steel producing country, with data on production, imports and exports. The looseleaf pages are distributed to subscribers as and when the information is received and printed. *World metal statistics* (World Bureau of Metal Statistics, 1948-), issued monthly, is another publication which now contains information on all types of metals and has a world wide coverage, although until 1967 it covered a much narrower field. Two annual OECD publications are *The iron and steel industry* (OECD, 1954-) and *The non ferrous metals industry* (OECD, 1954-), both covering member countries. The United Nations' Economic Commission for Europe issues a *Quarterly bulletin of steel statistics for Europe* (UN, 1950-) and the annual *Statistics of world trade in steel* (UN, 1913/59-), the latter devoted only to the foreign trade aspects. The Statistical Office of the European Communities publishes bi-monthly and annual issues of *Iron and steel* (EEC, 1962-) with data for the Common Market countries. Finally, the Commonwealth Economic Committee issued *Iron and steel and alloying metals* (HMSO, 1962-), presenting reviews on world wide consumption, resources, production, trade, stocks, and prices, which will, no doubt, be continued by the Commonwealth Secretariat.

TEXTILES

The Textile Council, until recently the Cotton Board, issues a *Quarterly statistical review* (the Council, 1946-) which has information on spinning, weaving, doubling, man made fibres, warp knitting, processed continuous filament yarns, finishing, production, imports and exports, etc, mainly for the United Kingdom but also including some world figures. Monthly tables of cotton production are published in the *Board of Trade journal*. An important US annual is the National Cotton Council of America's *Cotton counts its customers,* an annual statistical report on the quantities of cotton and competing materials consumed in the major textile products manufactured in the US. The International Federation of Cotton and Allied Textile Industries issue an annual *International cotton industry statistics* (the Federation, 1958-) and a bi-annual *European cotton industry statistics,* which is more limited geographically. Both have information on spinning, weaving, raw materials consumed, number of looms, cloth produced, hours worked. *Cotton: world statistics,* a quarterly, and *Cotton: monthly review of the world situation,* both published by the International Cotton Advisory Committee, give information on production, imports and exports of cotton yarn and piece goods in individual countries.

The Ministry of Commerce, Northern Ireland, issues a basic volume of *Review of the linen industry* (the ministry, 1955-) every few years with annual supplements to update the basic volume. Data includes statistics of raw materials, production, imports and exports.

The analysis of production of man made fibres is published monthly and quarterly in the *Board of Trade journal* and international data can be obtained from the International Rayon and Synthetic Fibres Committee (CIRFS).

The Wool Industry Bureau of Statistics publish a *Monthly bulletin of statistics* (the Bureau, 1950-) containing detailed data on the industry, including stocks, production, consumption, deliveries, machine activity and personnel. Monthly tables of wool exports are published in the *Board of Trade journal.*

Textil organon (Textile Economics Bureau Inc, 1930-) is a monthly periodical, with occasional 'basic book issues', which contains world wide information on man made fibres and wool. On textiles generally, the Commonwealth Council's *Industrial fibres* and OECD's *Textile industry in the* OECD *countries* (OECD, 1953-) are both world wide reviews of the production, trade and consumption of textiles.

TIMBER

The Timber Trade Federation of the United Kingdom issues a number of statistical publications including a *Yearbook of timber statistics,* having data on production, stocks, consumption, wholesale price index numbers, imports and exports of timber, and will often furnish other more detailed information if required. The Food and Agriculture Organization's *Yearbook of forest products statistics* (FAO, 1947-), compiled in collaboration with the United Nations' Economic Commission for Europe, includes statistics for nearly 200 countries, giving detailed data on round wood, processed wood, and pulp and pulp products. The commission also publishes its own quarterly *Timber bulletin for Europe* (UN, 1948-) and OECD issues annually *Timber market in the* OECD *countries* (OECD, 1953-), on the production and trade in timber, and *Pulp and paper* (OECD, 1954-), a statistical report on the situation of the pulp and paper markets of Europe and North America.

MOTOR VEHICLES

The Society of Motor Manufacturers and Traders provides the main basic statistical information on motor vehicles in the United Kingdom. The annual *Motor industry of Great Britain* (SMMT, 1947-) contains detailed statistics of the industry, including production and registration, both for the United Kingdom and for overseas countries. United Kingdom overseas trade in motor vehicles is also covered. The *Monthly statistical review* updates the annual with more recent monthly figures but is confined to data for the United Kingdom. Statistics of car and commercial vehicle production and of motor trade turnover are published monthly in the *Board of Trade journal* and there are also a number of tables in the *Monthly digest of statistics*. The US Automobile Association's annual *Automobile facts and figures* and *Motor truck facts* are digests of statistical information, some of it unique to the association. The Economist Intelligence Unit issues a quarterly *Motor business* which is devoted to the problems of the international motor industry, but McGraw-Hill's *World automotive market survey* is perhaps the most important publication containing international statistics of the industry. Published annually, the 1966 edition contained the results of a world motor census, a marketing map for thirty nine countries showing vehicle registrations, US exports in 1966, international trade in motor vehicles, and free world manufacture and assembly of motor vehicles in 1965.

OTHER MANUFACTURING INDUSTRIES

There are many other statistical publications on particular industries and it would not be possible to conclude them all. The International Rubber Study Group's monthly *Rubber statistical bulletin* (the Group, 1947-) includes data on the production, stocks, imports and exports, consumption, and prices of natural and synthetic rubber and of the end products. The Furniture Development Council issues a number of publications including an annual *Economic review for the furniture industry,* which has many statistical tables, an annual digest of statistics dealing with the domestic furniture industries of all the European member countries of the Union

57

Européenne de l'Ameublement which is published as *The furniture industry in Western Europe*, and a *Monthly bulletin of furniture statistics*.

The National Economic Development Office surveyed the statistics available for the electronics industry and the results, together with discussions on the pros and cons for improvements, were issued by the office as *Statistics of the electronics industry* (NEDO, 1967).

Statistical information on chemicals is not very plentiful and the Organization for Economic Cooperation and Development's annual *The chemical industry* (OECD, 1953-) is, therefore, a valuable source of reference. *Fertilizers* (OECD, 1951/54-) is also useful.

CONSTRUCTION

Just published is a *Directory of construction statistics*, compiled by Cynthia Cockburn and Michael Verstage for the Ministry of Public Buildings and Works (HMSO, 1968) which gives the results of a survey of existing statistics of the construction industry of Great Britain and includes a guide to the range of the statistics generally available. A more detailed *Inventory of construction statistics*, aimed at the specialist and concerned with the detailed analysis of statistical material, has also been issued but is not generally on sale, although copies have been deposited in certain libraries, a list of which can be obtained from the ministry.

The ministry currently issues a *Monthly bulletin of construction statistics* (MPBW, 1947-) and there is also an annual issue. These publications include cost and production indices, value of output and new orders, employment in the industry, housing (completed, approved, started, etc), production of materials and components, exports of building fittings and materials, industrial building, and building control. An annual *Sand and gravel production* (HMSO, 1954-) is also compiled by the ministry.

In the United States there is a *Source book of statistics relating to construction*, by Robert E Lipsey and Doris Preston (National Bureau of Economic Research/Columbia University Press, 1966). The product of many years of data collecting, this volume covers construction expenditure, contracts, permits and starts fairly completely and gives a sampling of the main series on construction materials. More up to date data is given in the monthly *Construction reports*, issued by the Bureau of the Census, and by Business and Defence Services Administration's *Construction review* (GPO, 1955-) which is issued monthly with an occasional supplement such as the *Construction statistics, 1915-1964*, which was published in 1966.

Construction statistics (UN, 1965), issued by the United Nations' Department of Economic Affairs as ' Studies in methods, Series F, No 13 ', discusses the problems encountered and the purposes of construction statistics. It indicates the type of data now being published in selected countries and the type of data required, focusing particularly on methods and techniques. Factual statistics are given in the United Nations' Economic Commission for Europe's *Quarterly housing construction summary for Europe* (UN, 1963-) which replaced the *Quarterly bulletin of housing and building statistics for Europe* and contains data on houses started, houses under construction and houses completed in individual European countries (see also the section on housing statistics in chapter three).

CHAPTER SEVEN

Trade

DISTRIBUTION. Distribution statistics deal with 'consumer goods', those which pass through the wholesale and retail trades to the consumer. The goods may move only through a retailer on their way from the producer to the consumer but it is much more likely that they will go through a number of hands before reaching their destination. Associated services such as hire, repair, catering and hairdressing are also involved.

Perhaps because of its complexity, published statistics of distribution are rather scarce. The Hopkins Committee (*Report of the Census of Distribution Committee, 1946*) recommended that a census of distribution should be taken in Great Britain and after a pilot survey taken in 1947, for which no results were published, the first census was taken for 1950. The census covered all establishments engaged in selling goods and certain service trades such as caterers, hairdressers, shoe repairers, etc. The results of the census were published in four volumes as *Census of distribution and other services, 1950*. The first volume to be published was a short report on the retail trade and then came volume I, Retail and service trades, area tables; volume II, Retail and service trades, general tables; and volume III, wholesale trades. Another volume, *Britain's shops: a statistical summary of shops and service establishments,* whilst not strictly a part of the census, analyses the register of establishments by main types of goods sold, and by county, metropolitan borough, and main town. The 1950 census was the only one to cover the wholesale trade as well as retail and service trades and volume III includes data on the number of wholesale establishments, kind of business, sales, employment, and employees' remuneration for Great Britain as a whole.

In 1954 the Verdon Smith Committee reported on the censuses of production and distribution and recommended that censuses of distribution be taken every ten years. Consequently, another full census was taken for 1961 and the results published in fourteen separate parts, most of which are devoted to one particular geographical area, although there are also three more general volumes of establishment tables, a summary of area tables, and organisation tables. Each area volume contains statistical data on the number of establishments, turnover, persons engaged, branches of retail establishments, and the number of retail outlets selling certain commodities. Although not recommended by the Verdon Smith Committee, less detailed censuses were taken for the years 1957 and 1966.

Analysis of the detailed information collected for the censuses takes time, but preliminary results are always published in the *Board of Trade journal* as soon as they are available, the first results of the 1966 census, for instance, were published in the journal for 23rd February 1968.

Large scale enquiries relating to wholesaling, catering and the motor trade were first taken in 1959, 1960 and 1962 respectively and are each to be repeated at intervals of five years. Information includes statistics of sales, stocks and fixed capital expenditure, and the results are published in the *Board of Trade journal*. An annual table on stocks and capital expenditure in the distribution and service trades, quarterly tables of manufacturers', wholesalers' and distributors' stocks, and a monthly table of retailers' stocks are all published in the journal, also.

Short term indications of changes in the pattern of retail trade are shown by the monthly tables of retail trade index numbers published in the *Board of Trade journal,* the more important figures also appearing in the *Monthly digest of statistics* and the *Annual abstract of statistics.* Also in the journal are monthly tables showing catering trade turnover, laundries and dry cleaners turnover, hairdressing trades turnover, and hire purchase business.

The United States *Census of business* (GPO, 1929-) is now taken every five years by the Bureau of the Census, covering wholesale, retail and service trades, numbers of establishments, employment and payroll, and receipts. After each census preliminary results are issued in leaflet or booklet form as the information becomes available, then preprints of the numerous parts of the final reports, and later the final bound volumes. The monthly *Survey of current business* (GPO, 1921-) compiled by the Office of Business Economics, includes statistical data on retail and wholesale sales, and there are also the *Current retail trade reports,* issued by the Bureau of the Census, comprising several different titles including *Monthly retail trade.* The revised edition of the United Nations Department of Economic and Social Affairs' *Bibliography of industrial and distributive-trade statistics* (UN, 1965), lists information being collected by each country and the publications in which the information appears. The only distribution statistics to be published on an international basis are those included in the general compilations of the international organisations and are referred to in chapter one.

FOREIGN TRADE

The records of the United Kingdom overseas trade go back to the thirteenth century when quantities and values of goods loaded and unloaded at the ports were recorded on parchment rolls. Later, from 1428 to 1799, the information was recorded in ' Port books'. Not really suitable to be used for the compilation of national overseas trade statistics, they have been used together with other source material to provide some interesting works such as G N Clark's *Guide to English commercial statistics, 1696-1782* (Royal Historical Society, 1938), which has an interesting discourse on the early trade statistics and also a bibliography of sources, Elizabeth Schumpeter's *English overseas trade statistics, 1697-1808* (OUP, 1960), and W Schlote's *British overseas trade from 1700 to the 1930's* (Blackwell, 1952). Some early trade statistics are also to be found in more general compilations, such as D Macpherson's *Annals of commerce* (1805), J Marshall's *A digest of all the accounts . . of the United Kingdom . . .* (1834), and G R Porter's *The progress of the nation* (second edition 1847).

Official statistics of foreign trade are collected by HM Customs and Excise, who maintain the accounts under about three thousand commodity headings separately distinguished in the annual *Export list* (HMSO) and the *Statistical classification for imported goods and for reexported goods* (HMSO). The order and grouping of the headings is based on the United Nations' *Standard international trade classification, revised* which, in turn, corresponds to the *Brussels tariff nomenclature,* used by many European countries. Monthly statistics are published approximately three weeks after the end of each month in *Overseas trade accounts of the United Kingdom* (HMSO, 1965-), which superseded the *Accounts relating to the trade and navigation of the United Kingdom* (HMSO, 1848-1964) in January 1965. The current title includes general summary tables of foreign trade, followed by detailed statistics of imports, exports and re-exports arranged by commodity and further subdivided by the main countries of origin and destination. The value and the quantity of the merchandise is shown for the month, for the cumulated months of the year to date, and the corresponding cumulation for the previous year. There are also tables showing statistics of goods liable to revenue duties, and finally a detailed index of commodities. Information in amplification of these accounts and in advance of the more detailed annual compilation (see below) can be obtained regularly or on an *ad hoc* basis from The Controller, Statistical Office of Customs and Excise, 27 Victoria Avenue, Southend on Sea, Essex, on payment of a fee to cover the cost of preparing the information. Perhaps it should be explained here that although the accounts are presented to Parliament by the President of the Board of Trade, the statistical information is collected and analysed by HM Customs and Excise.

Also published is the more detailed *Annual statement of the trade of the United Kingdom with commonwealth countries and foreign countries* (HMSO, 1853-). From the 1963 edition each issue is published in five volumes, having previously for many years appeared in four volumes plus an annual and a three yearly supplement. The first volume is devoted to summary tables of imports, exports and re-exports which give an overall picture of the foreign trade of the country; the second and third volumes contain detailed statistics of imports and re-exports arranged by the commodity classification, subdivided by the main countries of origin, and statistics of exports arranged by the commodity classification, subdivided by the countries of destination respectively; the fourth volume contains statistics of trade by countries, subdivided by commodities; and the fifth volume, which used to be published only every third year as a supplement to volume four, deals with trade through the various sea and air ports of the United Kingdom. A separate volume of *Protective duties* contains data on the values or quantities of imported goods entered for home use in the United Kingdom and receipts of duty. The installation of a computer at the offices of HM Customs and Excise at Southend has speeded up the preparation of the statistics considerably and it is expected that publication of the detailed *Annual statement . . .*

will eventually take place some two or three months after the end of the year covered.

As with production statistics, the need to ensure the confidentiality of statistics supplied by individual firms results in the available information sometimes being less detailed than it otherwise might have been.

Important features of changes in United Kingdom foreign trade are shown in the Board of Trade's monthly *Report on overseas trade* (HMSO, 1950-) which summarises, in tabular form, the country's trade with the rest of the world. One or two more general tables of foreign trade statistics appear first in the weekly *Board of Trade journal* and export figures for particular industries are often included in the board's *Business monitor: production series.* *Exports of works of art . . . report of the reviewing committee* (HMSO, 1953-), an annual report issued by the Department of Education and Science, contains statistical data on the licences issued and exports and imports of works of art. Trade associations and other organisations sometimes publish bulletins of foreign trade statistics for the products in which they are interested, such as the National Wool Textile Export Corporation's quarterly *Statistical bulletin,* the British Footwear Manufacturers Federation's *Footwear statistics—UK exports and imports,* and the Timber Trade Federation's *Analysis of UK timber imports by port of entry* and *Summary of imports and exports of sawn and planed softwood.* Many similar publications also include information other than foreign trade statistics and some of these have been mentioned in chapter six.

Very little is published on invisible trade, which includes payment for services, interests, profits, dividends and transfers of remittances and grants, but statistics are included in the ' pink book ', *United Kingdom balance of payments,* which is referred to in more detail later in this chapter. The British National Export Council issued a monograph in 1967 on invisible exports, *Britain's invisible earnings: the report of the Committee on Invisible Exports* (BNEC, 1967) which contains some statistical tables, mainly from published sources.

Although a part of the United Kingdom, Jersey publishes its own annual volume of foreign trade statistics, *Statistics of imports into, and exports and re-exports from the island of Jersey,* which is published by the States of Jersey.

The Ministry of Commerce, Northern Ireland compiles and publishes an annual *Summary of the trade of Northern Ireland* (the ministry), each issue having statistics for three separate years on balance of trade, imports and exports by classes and groups of commodities, trade at the principal ports, trade by countries, details of imports and exports classified by commodities, and trade between Northern Ireland and the Irish Republic by land.

The Commonwealth and sterling area: statistical abstract (HMSO, 1850-) compiled annually by the Board of Trade, summarises the external trade of the Commonwealth and sterling area countries individually and as a group. Apart from data on foreign trade, the abstract includes some information on prices of commodities, production, consumption and population. A number of the tables covering the foreign trade of overseas sterling countries are continued by later figures in a quarterly article in the *Board of Trade journal*, ' Overseas sterling area trade summary'.

Foreign commerce and navigation of the United States (GPO, 1821-) was prepared annually by the Bureau of the Census from 1821 to 1946. It was latterly published in three volumes with a fourth volume every other year, and contained similar information for the United States to that given in the *Annual statement of trade*... for the United Kingdom. It was lately decided that the publication should be revived with a volume for the year 1964, and a summary volume *Foreign commerce and navigation of the United States, 1946-1963* (GPO, 1965) has been issued to bridge the gap. Monthly statistics of foreign trade are published in a series of FT (foreign trade) reports. The series numbers, titles and contents of these reports have changed several times in the past but the most important current titles are FT *135 US imports, general and consumption, schedule A, commodity by country* and FT *410 US exports, schedule B, commodity by country*, schedules A and B being the classification schedules for imports and exports respectively. A useful guide to all current FT reports is *Guide to foreign trade statistics: 1967*, which is obtainable from the Bureau of the Census.

Practically all the countries in the world issue some foreign trade statistics and GATT, the General Agreement on Tariffs and Trade, published last year a *Compendium of sources: international trade statistics*, which lists the foreign trade statistics published by international agencies and national governments. *International*

65

trade statistics, by R G D Allen and J Edward Ely (Chapman & Hall, 1953), although published fifteen years ago is still of some value to those dealing with this type of material although the information it contains on titles of publications is very much out of date.

Perhaps the most important publication dealing with foreign trade statistics internationally is the United Nations Statistical Office's *Commodity trade statistics* (UN, 1952-), which is issued quarterly in a number of fascicules published as the information arrives from the individual reporting countries. Arrangement is by country, subdivided by commodities (using the *Standard international trade classification*) and then by country of origin for imports and destination for exports. Before 1962 the arrangement was primarily by commodity but in order to publish the data more speedily it was decided to publish the returns for each country as they were received. However, by agreement with the UN, Walker and Company use the computer tapes to compile *World trade annual,* which is arranged primarily by commodity, the trade of the industrialised nations being published in the four main volumes, whilst the trade of the industrialised nations with eastern Europe and the developing nations covers a further four supplementary volumes. The United Nations also publishes a *Yearbook of international trade statistics,* which is a much more general compilation of national tables. GATT also issues an annual *International trade* which, although mainly textual, contains some statistical tables.

The *Foreign trade statistical bulletins* (OECD, 1950-) of the Organization for Economic Cooperation and Development are issued in three series now whereas until about 1958 they were issued in four. Currently, series A is quarterly and contains statistics of overall trade of each of the OECD member countries; series B is also quarterly and shows trade by commodity groupings, analysed by the main regions; and series C, issued twice a year with an annual supplement, contains a detailed analysis by products. Among the publications of the Statistical Office of the European Communities are both monthly and quarterly titles containing foreign trade statistics of the six member countries. Since 1966 the quarterly publication has appeared in twelve separate parts each quarter, each part devoted to a particular group of commodities, with the overall title *Foreign trade: analytical tables* (NIMEXE). The title of the monthly publication is *Foreign trade: monthly statistics*. *EFTA trade* (EFTA Information Office, 1959/63-), an annual account of the trade of the member countries of the European Free Trade Association, includes a number of statistical tables, and the *Board of Trade journal* also carries a quarterly table of figures of EFTA trade. Also issued by EFTA is an *Annual review of agricultural trade*.

BALANCE OF PAYMENTS

Balance of trade figures were published in the *Statistical abstract for the United Kingdom* until 1936. Figures for 1936, 1937 and 1938 are in the *Board of Trade journal* for 23rd February 1939 and none were published during the war. White papers were issued six monthly between 1946 and 1957 and annually from 1958 to 1962 but these were superseded by the 'pink book', *United Kingdom balance of payments* (HMSO, 1946/57-), which is prepared annually by the Central Statistical Office in collaboration with other government departments and the Bank of England, and contains a much wider range of information. Quarterly estimates of the balance of payments are published in *Economic trends,* in the *Board of Trade journal,* and in *Financial statistics.* The Treasury also issues annual *Preliminary estimates of national income and balance of payments* (HMSO, 1952-).

United States balance of payments statistics are published in the *Federal reserve bulletin* (Federal Reserve System, 1915-) and in the *Survey of current business* (GPO, 1921-).

The International Monetary Fund's *Balance of payments yearbook* (IMF, 1948-) is an annual looseleaf volume issued monthly in sections as the information is received from the various countries. It covers the international transactions of more than eighty countries. *International financial statistics* (IMF, 1948-) is a monthly publication which includes data on balance of payments, and *Direction of trade,* a monthly supplement to *International financial statistics,* issued jointly by the International Monetary Fund and the International Bank for Reconstruction and Development (often called the World Bank), contains detailed trade by country information arranged in a uniform pattern, the values being given in US dollars. There are eleven monthly issues and an annual issue which contains retrospective figures.

67

In his article 'Statistics of advertising' in *The sources and nature of the statistics of the United Kingdom,* edited by M G Kendall (Oliver & Boyd, 1952-57), Mark Abrams refers to the impossibility of arriving at a completely satisfactory definition of 'advertising' and suggests that a reasonably comprehensive definition would be that advertising expenditure comprises all money spent in buying and using facilities in any medium of communication, either to convey information to a third party or to attempt to influence his views or behaviour.

Pre war there was little statistical information on advertising and in 1942 the Advertisers' Association proposed to the National Institute for Economic and Social Research that it should carry out an enquiry to collect information and analyse it. Nicholas Kaldor and Rodney Silverman undertook the task and in 1948 published *A statistical analysis of advertising expenditure and of the revenue of the press* (CUP, 1948) which covers the period 1934 to 1938 in detail and provides some estimated figures for the period 1920 to 1944. The association commissioned a new enquiry in 1948 and this resulted in Rodney Silverman's *Advertising expenditure in 1948* (the Association, 1951).

Current statistical information on advertising is given in the *Stat review* (Legion Information Services Ltd, 1933-). This publication began as a quarterly with the title *Statistical review of press advertising,* lately became the *Monthly statistical review of press, TV, outdoor and radio advertising* and in January 1968 divided into two publications, *Advertising: a monthly commentary on advertising, marketing, media and sales promotion* and *Stat review,* the information for the latter now being provided by computer. The periodical *Audit bureau of circulation* has circulation figures of newspapers and periodicals.

Readership surveys are carried out primarily for market research purposes. The *Hulton readership survey* was made annually between 1946 and 1956; the Institute of Practitioners in Advertising's *National readership survey* has been made at intervals since 1954, two of the latest results being *Tables of reading frequency for twenty-six publications (July 66-June 67)* (IPA, 1967) and *Readership of certain regional publications in the north eastern ITV area . . .* (IPA, 1967).

Both the institute and the Advertisers' Association collect statistical information on advertising and, although much of it is confidential, may be able to supply required information.

Information on advertising expenditure in the United States is in the periodical *Marketing communications*, known for many years as *Printers' ink*, and also in the *Survey of current business* and the *Census of business*.

CHAPTER EIGHT

Finance

Prior to the report of the Radcliffe Committee, *Report of the committee on the working of the monetary system* (HMSO, 1959) published financial statistics were sparse and such as were available were scattered in various publications. Two important serials which were introduced as a result of the Radcliffe report are *Financial statistics* and the *Bank of England quarterly bulletin*, the statistics in the latter being largely confined to banking. *Financial statistics* (HMSO, 1962-) is a monthly, prepared by the Central Statistical Office in collaboration with the statistical divisions of other government departments and the Bank of England to bring together the key financial and monetary statistics of the United Kingdom. Tables include financial data concerning the Exchequer and central government, local authorities, public corporations, banking and money supply, other financial institutions, companies, capital issues and stock exchange transactions, interest rates and security prices, and overseas finance.

In the United States the *Federal reserve bulletin* (Federal Reserve System, 1915-) is a monthly publication which has comprehensive data on economic and business developments pertaining to banking and finance. The statistical tables cover information on the Federal Reserve banks, department store trading, consumer credit, production indexes, and international finance. The monthly *Federal reserve chart book on financial and business statistics*, which is also published by the Federal Reserve System, contains charts on financial and other statistics and is supplemented by an annual historical supplement, which has long range charts. The monthly *Survey of current business* (GPO, 1921-) also contains some statistical series on national income, gross national product, personal and farm income, etc.

The International Monetary Fund's *International financial statistics* (IMF, 1948-), issued monthly, includes monetary and banking statistics, both internationally and for individual countries. The publication is described more fully later in this chapter.

NATIONAL ACCOUNTS

A descriptive article by John E G Utting, 'National income and selected statistics' in *The sources and nature of the statistics of the United Kingdom*, edited by M G Kendall (Oliver & Boyd, 1952-57) is particularly useful for the history of national income statistics from 1938 to 1956, no official figures being published earlier than 1938. In 1956 the Central Statistical Office produced *National income statistics: sources and methods* (HMSO, 1956) as 'Studies in official statistics, no 3', which describes in detail the basis and methods used in calculating the figures published in white papers and blue books on national income and expenditure and also on social accounts. The first four chapters give a general description of the statistics and the remainder of the volume consists of chapters dealing exclusively with tables covering particular aspects of the economy. It has been kept up to date by notes in the annual blue books on national income and expenditure, but a revised edition is now being prepared on similar lines but extended scope to cover the additional information now available and this should be issued during 1968.

The 'blue book' itself, *National income and expenditure* (HMSO, 1938/40-), which is compiled annually by the Central Statistical Office, contains much information on the British economy as a whole, including gross national product, national expenditure, personal income and expenditure, taxation, capital formation, etc. The Treasury issues an annual *Preliminary estimates of national income and balance of payments* (HMSO, 1952-).

Government accounts are published at various stages throughout the year. Firstly, as *Civil estimates* (HMSO) in which are published estimates of expenditure of the supply services (army, navy, air, civil and revenue departments) in the coming financial year. Separate *Defence estimates* are also published, and supplementary estimates may be issued during the financial year if required. All the estimates are put together in *Estimates . . . memorandum by the Financial Secretary to the Treasury* (HMSO) which is issued before the Budget. Directly after the Budget speech the *Financial statement* (HMSO, 1878/79-), the estimate of revenue and expenditure, is laid before the House of Commons by the Chancellor of the Exchequer. *Public income and expenditure* (HMSO, 1869/70-), published within a month of the close of the financial year, shows the amount collected and spent in that year. The *Finance accounts* (HMSO, 1880/81-) are published in the autumn and set out in considerable detail the main heads of revenue and expenditure of the Exchequer in the preceding financial year; they also show particulars of the national debt and other loan transactions, although these are also published in the Treasury's *National debt return* (HMSO) and *Loans from the consolidated fund* (HMSO, 1961-), the annual report on government lending and the capital financing requirements of the nationalised industries, and the Exchequer and Audit Department's *Consolidated fund, abstract account* (HMSO, 1898/9-). Details of government trading services are given in the Exchequer and Audit Department's *Trading accounts and balance sheets* (HMSO, 1920-), the accounts and balance sheets of trading and commercial services conducted by government departments. *British aid: statistics of official economic aid to developing countries* (HMSO, 1961-) is compiled annually by the Ministry of Overseas Development and includes data on direct aid and contributions to multilateral agencies.

The Ministry of Finance, Northern Ireland issues an annual *Public income and expenditure* (HMSO, Belfast) and *Finance accounts of Northern Ireland* (HMSO, Belfast), the latter including public income and expenditure.

National accounts statistics for the United States are published in the *Survey of current business,* which was referred to earlier in this chapter, and a retrospective supplement *The national income and product accounts of the United States, 1929-1965: statistical tables* (GPO, 1967) has been compiled by the US Business Economics Office. They are also in the *Economic report of the President to Congress.* The New York Tax Foundation's *Facts and figures on government finance* (the Foundation, 1941-) is issued biennially and contains a wealth of information on federal, state and local government organisation, expenditure, income, debt, employment, etc.

The United Nations Statistical Office issues a *Yearbook of national accounts statistics* (UN, 1957-) which continues the half yearly *Statistics of national income and expenditure* (UN, 1952-56) and has both international and national tables on gross net product, gross domestic product, distribution of national income, capital formation, private consumption expenditure, government revenue and expenditure, etc. The United Nations also published *National and per capita incomes of seventy countries in 1949* (UN, 1952) and *Per capita national product of fifty five countries, 1952-54* (UN, 1957) but the series has been discontinued, unfortunately. Financial statistics, including government finance, are also included in *International financial statistics*.

The Organization for Economic Cooperation and Development issued *Statistics of national product and expenditure, 1938 and 1947 to 1955* (OECD, 1957) and continued this with *Statistics of national accounts* (OECD, 1950/61-). An annual volume of *National accounts* is also issued as a supplement to the European Communities' *General statistical bulletin*.

Balance of payments statistics are dealt with in chapter seven.

LOCAL GOVERNMENT FINANCIAL STATISTICS

V J Oxley's *Local government financial statistics* (Institute of Public Administration, 1951) is a guide to published statistics relating to the finance of local authorities in England and Wales. More recently, W Barker's *Local government statistics* (IMTA, 1965) includes financial as well as other types of local government statistics.

The Ministry of Housing and Local Government issue *Local government financial statistics* (HMSO, 1934/35-), an annual summary of income and expenditure of local authorities in England and Wales under principal heads of service. Similar information is given in the Scottish Development Department's *Local authority financial returns* (HMSO, 1961-) and the Ministry of Health and Local Government, Northern Ireland's *Local authority financial returns*. The Society of County Treasurers' *Financial and general statistics of county councils* (SCT, 1955/56-) includes statistics of rates and rateable values, penny rate product, road mileages, capital expenditure, net debt, smallholdings, and government grants and this is supplemented by *Capital expenditure of county councils, England and Wales* (SCT, 1961/62-).

Each local authority issues a budget and this involves detailed statistics of rates and rating which are published in *Rates and rateable values* (HMSO, 1921-), produced by the ministry and the Welsh Office, *Rates and rateable values in Scotland* (HMSO, 1939-)

and *Local authority rate statistics, Northern Ireland* (HMSO, Belfast). The Institute of Municipal Treasurers and Accountants also publishes similar information.

TAXATION

The *Annual report of the Commissioners of Inland Revenue* (HMSO, 1857/58-) has a statistical appendix which now includes data on income tax, surtax, survey of personal incomes, corporation tax, profits and excess profits taxes, the financial operation of companies, capital gains tax, death duties, stamp duties and other duties, war damage, tithe redemption, and valuation. The *Report of the Commissioners of HM Customs and Excise* (HMSO, 1909/10-) has tables on purchase tax as well as on duties generally and specifically. Less detailed figures are also published in *Economic trends* and the *Monthly digest of statistics*.

The Internal Revenue Service of the US Treasury issues *Statistics of income: individual income tax returns, Statistics of income: corporation tax returns,* and *Statistics of income: US business tax returns* (GPO, 1915-) each of which is issued annually and contains comprehensive statistical data on the returns filed during the year.

BANKING

The most up to date banking statistics are found in the monthly *Financial statistics,* and the *Bank of England quarterly bulletin* (the Bank, 1961-) a more selective quarterly. But the bulletin is an important publication, containing valuable articles on aspects of banking and finance as well as a statistical annexe which has data on Exchequer and central government, banking, capital markets, external finance, gold prices, exchange rates, short term money rates and security yields.

The main source of banking statistics in the United States is the monthly *Federal reserve bulletin,* described earlier in this chapter. Internationally, the International Monetary Fund's *International financial statistics* (IMF, 1948-) includes international and national monetary and banking statistics. Financial data on the fund and also on the International Bank for Reconstruction and Development, are followed by international data on gold reserves, foreign exchange, exchange rates, prices of major world trade commodities, freight rates, price indices, interest rates, changes in the money supply and in the cost of living, and values of world trade all given in US dollars and so enabling country comparisons to be made. National tables include similar but more detailed information.

The Stock Exchange issues annually *Statistics relating to securities quoted on the London Stock Exchange* and *Interest and dividends upon securities quoted on the Stock Exchange.* Indices of general movements in security prices are published in newspapers and economic or financial periodicals. They include *The F T actuaries share index* published daily in the *Financial times,* and explained in *Guide to the F T actuaries share index* (St Clement's Press), the *Financial times industrial ordinary share index, The times stock exchange indices* published daily in *The times* and explained in *The times share index* (The Times, 1964), and the *Economist extel indicator,* which is explained in *The economist* for 27th October 1956. Services such as those offered by Extel Statistical Services Ltd and Moodie's Services Limited are also valuable sources of often confidential information. The former firm offers a *Daily statistics (card) service* which covers about 5,000 British companies whose shares are quoted on the Stock Exchange and gives up to date data on their financial structure, profit history, dividend record and subsidiary interests, and *Compstat: statistics (computer) services* which is, in fact, several services covering daily price ratio tables, lists of redemption yields and mid market closing prices of British funds and comparative market valuation and yield ratios of British equities. Moodie's also offer a card service, a weekly *Moodie's review of investment, Moodie's investment handbook* which is updated quarterly and covers about 1,000 companies, and *Moodie's investment digest,* a compact annual guide to leading shares.

Extel has separate card services for North American, Australian and European companies, and Moodie's has an overseas card service as well as being connected with Moody's Investors Service Inc (note the different spelling), who issue a number of publications including *Moody's handbook of common stocks,* covering United States companies. Standard and Poor's Corporation issues an annual *Trade and securities statistics* which reproduces the leading indices and also contains original ones of security prices and yields. It also includes some information on business conditions,

75

cost of living and prices, and is, in part, kept up to date by a monthly *Current statistics*.

Official statistics are given in two annual Board of Trade publications, *Life and other long term insurance business: statements deposited with the Board of Trade* (HMSO, 1871-), which is concerned with life assurance business, and *Insurance business: summary of statements deposited with the Board of Trade* (HMSO, 1949-), which is concerned with other types of insurance business. The former title has varied throughout the years and before 1949 contained some of the information now published in the latter title. Prior to 1939 the individual statements returned by insurance companies were given in full. Insurance companies transactions and pension funds transactions are published quarterly in the *Board of Trade journal*.

Statistics of industrial life assurance are included in the Registry of Friendly Societies annual *Report of the Industrial Assurance Commissioners* (HMSO, 1875-). Statistics of registrations, legal cases, benefits, assets, etc of the friendly societies are given in the registry's *Annual report of the Chief Registrar of Friendly Societies* (HMSO, 1875-) now issued in five parts: general, friendly societies, industrial and provident societies, trade unions, and building societies. Actuarial statistics are published periodically in the *Journal of the Institute of Actuaries* and in the *Transactions of the Faculty of Actuaries*. The Life Office's Association, the Associated Life Office's Association and the Industrial Life Office's Association issue an annual booklet *Life assurance in the United Kingdom* which includes statistics in respect of the assurance companies which are members of the associations.

The US Special Libraries Association has published *Sources of insurance statistics* (SLA, 1965), edited by Elizabeth Ferguson. It is a detailed index to insurance statistics published on a regular basis in the USA and Canada.

Companies general annual report (HMSO, 1891-) is compiled by the Board of Trade and contains figures of the number of companies on the register, new registrations and removals, private companies, liquidations, compulsory winding up, etc. Monthly statistics of new companies registered and of partnerships registered in Great Britain and also in Northern Ireland are published in the *Board of Trade journal,* as are quarterly tables on company income and finance, and on capital expenditure by industry, and an annual table on capital expenditure on the distributive service trades. Varying amounts of statistical information are given in the annual reports of individual companies.

Bankruptcy statistics are included in the board's *Bankruptcy general annual report* (HMSO, 1883-), *Bankruptcy and companies (winding up) proceedings account* (HMSO, 1926/27-), and in a quarterly table of bankruptcy orders published in the *Board of Trade journal.*

In 1955 the National Institute for Economic and Social Research produced *A classified list of large companies engaged in British industry* which ranged companies by the size of their net assets and income in 1953/54. The Board of Trade continued what had been a pilot project and issued every three years a publication which is currently titled *Company assets, income and finance* (HMSO, 1957-). The 1963 edition includes over 2,000 of the largest companies, ranked in order of value of net assets, which are quoted on the United Kingdom stock exchanges and are mainly engaged in the manufacturing, distribution, construction, transport, property and certain other services. *The times 300 leading companies in Britain and overseas,* published annually by *The times,* and *Fortune directory,* published by Fortune magazine, are similar publications covering a wider geographical area.

The Securities and Exchange Commission of the US Federal Trade Commission issue a *Quarterly financial report for manufacturing corporations* (GPO) which includes data on the rates of change in sales and profits, profits per dollar of sales, annual rates of profit on stock holders equity, and financial statements in ratio form and on dollar amounts.

COOPERATIVE TRADING

The Cooperative Union Ltd, a non trading federation of cooperative societies which acts for them in an advisory manner, publishes an annual *Cooperative statistics* (the Union, 1938-) including data on the number and size of societies, membership share capital, reserves, employees and wages, trade, etc. Part three of the *Annual*

report of the Chief Registrar of Friendly Societies (HMSO, 1875-)
also includes statistics of the operations of cooperative societies.

CONSUMERS' EXPENDITURE

Three publications issued in a series ' Studies in the national income
and expenditure of the United Kingdom ', published under the joint
auspices of the National Institute for Economic and Social Research
and the Department of Applied Economics, Cambridge, are concerned with prewar consumer expenditure. They are *Consumers'*
expenditure in the United Kingdom, 1900-1919 by A R Prest (CUP,
1954) and *The measurement of consumers' expenditure and*
behaviour in the United Kingdom, 1920-1938, volume I of which
is by Richard Stone (CUP, 1954) and volume II by Richard Stone
and D A Rowe (CUP, 1966). These volumes contain many statistics
on supplies, consumption, prices, etc of particular commodities,
and the earlier volume covers some years prior to 1900 but in less
detail. Figures appear in the ' blue books ' on national income and
expenditure and in the *Monthly digest of statistics,* but the main
sources of information today are the Ministry of Labour's *Family*
expenditure survey (HMSO, 1957/65-) and the National Food Survey
Committee's *Domestic food consumption and expenditure* (HMSO,
1950-). The former is compiled chiefly to provide information
which enables the individual items of consumption included in
the ' Index of retail prices ' to be given their proper weights, but it
is also a very comprehensive source of information about the
incomes and expenditures of private households. It includes data
on individuals by sex, age and working status, the earnings of
individuals, household characteristics, sources of household income,
and income and expenditure analysis. An article on the increased
scope of the survey is in the *Ministry of Labour gazette,* January
1967. *Domestic food consumption and expenditure* is limited to
statistics of food consumption, expenditure and nutrition of private
households in Great Britain.

More specialised are two publications of the Tobacco Research
Council, *Statistics of smoking in the United Kingdom,* the fourth
edition being published in 1966, and *Tobacco consumption in*
various countries, published in 1963, both edited by G F Todd.
The *Report of the Commissioners of HM Customs and Excise,*
in its statistical tables on duties collected gives some indication

of the consumption on spirits, beer, wine, matches, mechanical lighters, etc.

The United States' Department of Agriculture's Economic Research Service has published *US food consumption; sources of data and trends, 1909-1963* (GPO, 1965) which includes data on per capita consumption, nutritive value of food available for consumption, supply and use of food, supply and use of farm commodities, prices, expenditure and population. There is also a chapter on food consumption surveys and a bibliography of the literature cited. Currently, *Current population reports: series P65, consumer buying indicators,* issued at irregular intervals by the US Department of Commerce, and *series P60, Consumer income* contain useful information.

In 1967 the International Labour Office issued *Household income and expenditure statistics. No 1, 1950-1964* which contains the principal results of household income and expenditure (budget) surveys in various countries, providing information in standardised form on levels of living, including sources of household income, distribution of household expenditure, distribution of consumption expenditure, distribution and expenditure on miscellaneous items and quantities consumed of major food items. Supplements are to be issued at intervals of one or two years covering the results of new household budget surveys. The Statistical Office of the European Communities issues a special series of economic accounts or household budgets for each of the member countries as a supplement to the series *Social statistics.*

CHAPTER NINE

Prices

Historical works include J E T Roger's *History of agriculture and prices, 1259-1793* (OUP, 1866-1903), T Tooke and W Newmarket's *History of prices, 1792-1856* (King, 1928), and the Board of Trade's *Wholesale and retail prices . . . in 1902, with comparative tables* (HMSO, 1903) which has some tables going back to 1810 and is concerned with specific commodities rather than prices generally. Long term price indices are also included in B R Mitchell and P Deane's *Abstract of British historical statistics* (CUP, 1962).

WHOLESALE PRICES

In 1949 the Board of Trade introduced a ' family of index numbers ' intended to be of use to industry, government and the economist. This new index was explained by J Stafford, the director of the board's Statistics Division, in an article ' Indices of wholesale prices ' which was published in the *Journal* of the Royal Statistical Society in 1951. A further article on the problems of obtaining and analysing the mass of information needed for the index is ' United Kingdom indices of wholesale prices ', by H S Phillips, and this was published in the same periodical in 1956. Other than these two articles, notes on the compilation of the index numbers are published each year in the *Board of Trade journal* with the annual review of price changes made each February. Index numbers of wholesale prices are published monthly in the *Board of Trade journal*, the *Monthly digest of statistics* and *Statistics on incomes, prices, employment and production*.

Average prices of selected commodities are published annually in the *Board of Trade journal*, and *Annual abstract of statistics*, and the *Journal* of the Royal Statistical Society, which includes each year a long run of average prices of forty five selected items

80

used in the construction of the Sauerbeck or Statist index of wholesale prices. Commodity prices also appear in the daily *Public ledger* and in the quarterly *National Institute economic review*. The *Daily commercial report* published by Bagot and Thompson Ltd is a series of closely printed cards giving a detailed listing of current prices and is issued as a service. Trade papers such as the *Metal bulletin* are often the source of prices of the commodities in which their subscribers are interested, and as these sources are so scattered and unpredictable the commercial departments of some of the larger public libraries keep their own indexes of sources of prices.

Agricultural price indices are published in the *Monthly digest of statistics* and the *Annual abstract of statistics;* the Ministry of Agriculture's weekly *Agricultural market report* covers prices at various markets for a large number of products; and the annual *Agricultural statistics* includes prices of cereals, farm crops and livestock.

RETAIL PRICES

The Ministry of Labour compiles the retail price index which indicates the standard and cost of living. An interim retail price index replaced an earlier cost of living index in 1947, and this was in turn replaced by the present index in 1956. The index, which is based on the actual average retail prices of basic foodstuffs and other items, is described in *Method of construction and calculation of the index of retail prices* (HMSO, *third edition* 1964), compiled by the Central Statistical Office and issued as number six in their 'Studies in official statistics'. The index is published monthly in the *Ministry of Labour gazette* and also, in less detail, in the *Board of Trade journal* and *Statistics on incomes, prices, employment and production*. Retail price indices of food are included in the annual report of the National Food Survey Committee, *Domestic food consumption and expenditure,* and in the *Family expenditure survey*, both of which are described in the section on consumers' expenditure in chapter eight.

A consumers' price index is included in the 'blue book', *National income and expenditure,* as is described in *National income: sources and methods,* both publications being described in more detail in chapter eight.

Indices of prices of imports and exports are published monthly in the *Board of Trade journal* and in *Statistics on incomes, prices, employment and production,* and annually in the *Annual abstract of statistics.*

The US *Labor Department's Bureau of Labor Statistics* compiles a monthly *Wholesale prices and price indices* (BLS, 1952-), indicating the changes in primary market prices of various commodities. The bureau also issues a monthly *Consumer price index* (BLS, 1953-) which indicates the changes in prices of goods and services. *Prices: a chartbook, 1953-62,* published as 'BLS bulletin 1351', with a supplement to September 1963, is a compilation based mainly on data from other publications and covering wholesale and consumer prices and price indices.

There is no guide to United Kingdom prices comparable to Paul Wasserman's *Sources of commodity prices* (ALA, 1960) which gives American and a few Canadian sources of prices for over 6,800 products and commodities, together with a note on the market or markets wherein the prices are effective. Published in 1960, however, a new edition would seem to be needed.

International wholesale price indices are published in *International financial statistics,* and the United Nations' *Monthly bulletin of statistics* and *Statistical yearbook.* Retail price indices are in *International financial statistics,* and the International Labour Office's *Bulletin of labour statistics* and *Yearbook of labour statistics.* A survey of living costs in New York as compared with certain capitals in the world is made by the United Nations. First published as a monograph in 1952 and revised in 1954, 1959 and 1962 in the series M of the United Nations statistical papers, it is now published twice a year, in the April and October issues of the *Monthly bulletin of statistics* under the title 'Retail price comparisons to determine salary differentials of United Nations officials'.

The International Trade Centre of the General Agreement on Tariffs and Trade (GATT) produced a *Compendium of sources: basic commodity statistics* in 1967 which includes basic sources of prices of primary commodities, as well as sources of production, consumption and trade statistics.

CHAPTER TEN

Transport and Communications

In *Statistics for economists* (Hutchinson, second edition 1952) R G D Allen wrote 'statistics on transport and communications are voluminous for railways, adequate for shipping and the post office, recently much improved for civil aviation and very scanty for road transport'. The situation has improved somewhat since 1952 and is still improving but, although there has for a long time been some sort of public control over the operation of transport the mass of statistical information collected is scattered over a wide range of sources and publications. Nowhere is it brought together in a comprehensive publication concerning statistics of transport as a whole. In November 1967 the ASLIB Transport Group held a conference on statistics in transport, the aim of which was to review the statistical information available, published and unpublished sources, their availablity to research workers and members of the public, the purposes for which they are compiled, and how they are used. The proceedings of the conference are to be published later in 1968.

An article by F A A Menzler, 'Rail and road', published in *The sources and nature of the statistics of the United Kingdom,* edited by M G Kendall (Oliver & Boyd, 1952-57), describes in detail the earlier statistical series on railways and roads. In the same compilation is an article by M G Kendall on 'Merchant shipping' and one by A Major Lester on 'International air transport statistics', the latter concerned mainly with the setting up of the International Civil Aviation Organisation and its work in collecting statistics. More recent information on publications dealing with road and passenger transport statistics from the local government angle is in W Barker's *Local government statistics* (IMTA, 1965). *Transportation: information sources* (Gale, 1965), by Ken N Metcalf, devotes a chapter to the statistics of US transport.

83

The Ministry of Transport's *Passenger transport in Great Britain* (HMSO, 1962-) presents statistics of passenger movement in Great Britain, by all the main forms of inland transport, by rail, by public road transport, by private road vehicles, and by air. Formerly it was issued as *Public road passenger statistics* (HMSO, 1949-62) with a more limited coverage. The *Annual report and accounts of the London Transport Board* (HMSO) has statistics of capital assets, permanent way, and rolling stock of the underground railways and assets and operations of road transport undertakings in the area covered by the board. Long runs of summary statistics of all kinds of transport are in the *Monthly digest of statistics* and the *Annual abstract of statistics*. A *National travel survey* was undertaken by the Ministry of Transport in 1964, but so far only a preliminary report has been issued by the ministry. This is in two parts, one dealing with household vehicle ownership and use and the other with personal travel by public and private transport.

The United Nations Economic Commission for Europe publishes an *Annual bulletin of transport statistics for Europe* (UN, 1949-) which includes data for each European country on passenger transport; freight transport by rail, road and inland waterway; railway, road, waterway and pipeline networks; mobile equipment; and the utilisation of equipment.

RAILWAYS

Railway returns (HMSO) were compiled annually, first by the Board of Trade and later by the Ministry of Transport, from 1838 to 1938, except for the period of the first world war. This publication contained the detailed financial, traffic and operating statistics of the various railway companies. *Railway statistics* (HMSO), issued monthly by the Ministry of Transport from 1920 to 1939, contained statistics of traffic and operating results on either a monthly or a four weekly basis. Both publications ceased with the outbreak of war but the ministry later published summary annual returns for the periods 1938-44, 1945, 1946 and 1947 to fill the gap between prewar titles and the British Transport Commission's publications. Mention must also be made of the statistical information contained in the annual reports of the London Passenger Transport Board from 1933 to 1947.

After the nationalisation of the railways on 1st January 1948, the British Transport Commission produced an annual report which, in effect, was a continuation of *Railway returns* and included statistical data on rolling stock, locomotives, coaching and freight vehicles, length and state of track, and capital equipment as well as information on roads. From 1963 the information concerning railways has been continued in the *Annual report and accounts of the British Railways Board* (HMSO). From 1948 the commission also published a monthly *Transport statistics* (BTC, 1948-62), continuing the information given in the prewar *Railway statistics* as well as including data on road transport and inland waterways. This has been continued in part by the board's four weekly *Statistics* (BRB, 1963-). Finally, the board also publishes a popular booklet, *Facts and figures about British rail*, the 1967 edition having statistical data for the years 1948 to 1966 on many aspects of British railways.

Statistics of railway accidents were published in *Railway accident returns* (HMSO, 1840-1938) and since the war in the annual *Report to the Minister of Transport on the accidents which occurred on the railways of Great Britain* (HMSO).

The principal publication in the United States is the Interstate Commerce Commission's *Transport statistics in the United States* (GPO, 1887/88-), currently issued in six parts (numbered 1 to 9, parts 2-4 having been absorbed by other parts during the years). Popularly called the ' blue book ' it covers railroads, sleeping car companies, express companies, electric railways, water carriers, oil pipe lines, motor carriers, freight forwarders, and private car lines, and it includes data on mileage, equipment, financial and operating statistics. Other organisations such as the Bureau of Transport Economics and Statistics and the Association of American Railroads also publish statistics, but they are based mainly on the same source material as that used by the commission.

The most important source of statistical information on international railway matters is in the International Union of Railways' *International railway statistics* (the Union, 1925-), which has been published annually since 1925 with a break between 1938 and 1946. The tables cover a wide field of data on the track, locomotives, operations, financial information, accidents, staff, etc. Before the war a monthly as well as an annual issue was published.

Shipping statistics are of two kinds: size and type of ships in the fleet, and operations, *ie* goods and passengers carried. The most detailed statistics on ships are published by Lloyd's Register of Shipping but as these give world coverage they are referred to in more detail later in this chapter. Giving similar coverage, but mainly concerned with United Kingdom shipping, is the *Annual report of the Chamber of Shipping of Great Britain*. Statistical tables include an analysis of ships of the fleet by size, age, and type, ships' tonnage, shipbuilding, ships lost, shipping movement, freights, and laid up shipping. River and estuarial craft, fishing vessels and ships under 100 gross tons are excluded. Statistical information on fishing vessels is given in *Sea fisheries statistical tables, Scottish sea fisheries statistical tables,* the *Annual report of the White Fish Authority,* and the *Annual report of the Herring Industry Board,* all of which are referred to in more detail in chapter six. The Board of Trade issues an annual return of *Shipping casualties and deaths* (HMSO) relating to vessels registered in the United Kingdom. Needless to say the publications mentioned above do not include data on ships of the Royal Navy.

The Rochdale Committee, in its *Report of the committee of enquiry into the major ports of Great Britain* (HMSO, 1962), pointed out that there was no comprehensive source of port statistics. Since 1964, therefore, the National Ports Council has published an annual *Digest of port statistics* (HMSO, 1964-) with a view to providing in one publication the basic statistical information covering the main features of and activities in British ports. The digest includes data on port organisation and harbour operations, capital expenditure, labour, goods traffic by commodity analysis and by overseas trading areas, passenger traffic and shipping movements. The statistics are derived from information provided at the special request of the council by port authorities, government departments, and non government organisations. Prior to 1964 one had to rely on the *Annual report of the British Transport Docks Board* (HMSO, 1963-), which includes comparative figures for facilities, traffic handled and finances of the twenty six ports operated by the board, and on the reports of individual boards. The Port of London Authority, for instance, issues an *Annual report and accounts* (PLA, 1910-) which contains a section of charts and statistics of the goods trade of the port, shipping, goods handled at the quays, warehouse stocks, manpower, and financial data, as well as an annual popular booklet, *The Port of London Authority: some facts and figures*. A monthly statistical table of shipping movements at United Kingdom ports and an annual table showing the nationality of vessels in United Kingdom seaborne trade are published in the *Board of Trade journal*.

Tramp freight rates are issued to the press monthly by the Chamber of Shipping of the United Kingdom and are also published in the *Monthly digest of statistics*.

Detailed statistics of trade through the principal ports of the United Kingdom and of goods transhipped at principal UK sea and air ports are published in volume V of HM Customs and Excise's *Annual statement of trade of the United Kingdom with Commonwealth and foreign countries* (HMSO, 1853-). This volume now appears annually but prior to 1964 the tables it contains were issued only every three years as volume IV, Supplement (further details of this publication are given in chapter seven). *Port statistics for the foreign trade of the United Kingdom* (the Association, 1961-) is issued by the Dock and Harbour Authorities' Association annually in three volumes. Part I contains data on imports and exports at the twenty principal ports for each division of external trade; Part II, selected commodities imported or exported at the twenty principal ports of the United Kingdom together with the ten principal countries concerned; and Part III, imports and exports at individual ports for each division of external trade. For other publications showing the quantity and value of foreign trade, see chapter seven.

Statistics for inland waterways of the United Kingdom can be found in the *Annual report of the British Waterways Board* (HMSO, 1963-) which replaced, in part, the *Annual report and accounts of the British Transport Commission*.

The Marine Administration of the US Department of Commerce issues a number of publications dealing with shipping. The *Annual report* of the administration includes various statistical studies on subjects such as new ship construction, merchant ship deliveries, and the employment of merchant ships; *Employment report of United States flag merchant fleet seagoing vessels of one thousand gross tons and over* is a quarterly giving information on ownership, status, and area of employment of US ships. *Vessel entrances and clearances* (GPO) issued annually by the Bureau of the Census includes the number and net registered tonnage of vessels entering and clearing US customs areas. The Marine Administration issues *United States and Canada Great Lakes fleets* annually, and the American Waterways Operators publish mileage and traffic statistics in their annual *Inland waterborne commerce statistics*.

As already mentioned, Lloyd's Register of Shipping publish several statistical serials of world shipping registered at Lloyd's, including the annual *Statistical tables,* which has data on the country of registration, size and age, type, and propulsion, and also giving numbers and tonnage of ships registered, launched, and lost. Other titles published are the *Annual summary of merchant ships launched in the world,* the *Statistical summary of merchant ships totally lost, broken up, etc,* the quarterly *Merchant shipbuilding return* and *Casualty return.* Lloyd's have published these statistical serials for many years, suspending their publication during the second world war; they will often supply unpublished information but this depends entirely upon the actual nature of the enquiry. The Marine Administration of the US Department of Commerce also issues international statistics of shipping, including the annual *Merchant fleets of the world, Merchant ships built in the United States and other countries, New ship construction, New ship deliveries,* and the semi annual *A statistical analysis of the world's merchant fleets,* showing the number and tonnage of ships, distribution by types, new construction, transfers and losses by scrappings. Fearnley & Egers Chartering Co Ltd now issue an annual *Trades of world bulk carriers* (1960-) which aims to illustrate the development in international sea transport of the main bulk commodities as well as the participation of bulk carriers in these trades.

Maritime transport (OECD, 1962-), published annually by the Organization for Economic Cooperation and Development in the 'Trends in economic sectors' series, is a report on shipping and shipbuilding in OECD member countries and has many statistical tables.

AIR

Air transport is a comparatively young member of the transport industry and consequently statistics of air transport are still very much in their infancy. The situation cannot have been improved by the transfer of responsibility from one government department to another, from the Air Ministry to the Ministry of Civil Aviation, which later became a part of the Ministry of Transport and Civil Aviation, broke away again, and is now a part of the Board of Trade. Previously issued by the Ministry of Civil Aviation and then by the Board of Trade were *Operating and traffic statistics of United Kingdom airlines,* issued quarterly and annually (both financial and calendar years), containing statistics of scheduled services, non scheduled services, inclusive tours and other separate fare charters, etc for each operating company, statistics of passengers, cargo and vehicle ferry services being shown separately. Tables showing aircraft type and utilisation and operations by type of licence are also included. Other titles were *Operating and traffic statistics of United Kingdom airways corporations and private companies,* issued monthly and annually, *Summary of activity at aerodromes in the United Kingdom and Channel Islands,* issued monthly with an annual summary, *National air traffic control service: summary of aircraft movements controlled by air traffic control centres,* issued monthly, and *Air transport movements diverted to Board of Trade aerodromes,* also monthly. All these publications had a controlled circulation and are available only from the Board of Trade, but as from the beginning of 1968 the information is being incorporated into a *Business monitor: civil aviation series.* There are at present five titles in the series, the first four to be published monthly with annual summaries and the fifth quarterly with an annual summary. They are *Airport activity,* containing details of numbers of landings and takeoffs on various kinds of flights at each of the important airports in the United Kingdom and the Channel Islands, *Air passengers,* having statistics of the terminal and transit passengers at airports, *Air freight and mail,* containing statistical data on cargoes, *Airlines,* having data on mileage flown, passengers, freight and mail traffic, capacity and load factors of BOAC, BEA and private companies. Finally, the quarterly with the same title, *Airlines,* containing operating and traffic statistics with details of the traffic and capacity of each airline on international and domestic, scheduled and non scheduled services together with details of the airlines' fleets and their utilisation. Each of these series is available on subscription from HMSO. Tables relating to air transport in the *Board of Trade journal* include monthly information on activity at United Kingdom aerodromes and on mileage and traffic of aircraft, quarterly information on aircraft noise at Heathrow Airport, and six monthly information on trade by air. The annual reports and accounts of the British Airports Authority, the British European Airways, the British Overseas Airways Corporation, and the British Independent Air Transport Association all include statistical information on traffic handled by members of each organisation.

Accidents to aircraft on the British register (HMSO), previously compiled annually by the Ministry of Civil Aviation, is now compiled by the Board of Trade.

The United States Federal Aviation Agency issues an annual *Statistical handbook of civil aviation* which contains data on certification, production and exports, accidents, and scheduled air carrier operations. It is the standard summary of official statistical data on civil aviation activity in the US. The agency also issues *Federal airways air traffic activity,* a semi annual statistical guide to activities of the federal airways system, *Air commerce traffic pattern* and *Air commerce traffic pattern, United States flag carriers,* respectively a semi annual on domestic flights and an annual on international flights, both giving statistics of aircraft departures, passengers, mail, and cargo. The Civil Aeronautics Board issues a number of statistical publications including an annual *Handbook of airline statistics,* containing statistics of US certificated air carriers, *Monthly air carrier traffic statistics,* quarterly *Air carrier financial statistics,* and an annual *Statistical review, United States air carrier accidents.* The Air Transport Association of America issues an annual statistical review of American aviation, *Facts and figures about air transportation,* and the Aerospace Industries Association of America issues an annual statistical source book, *Aerospace facts and figures,* which aims to bring information from various sources together in one volume.

The International Air Transport Association's annual *World air transport statistics* contains operating data of IATA members. The International Civil Aviation Organisation issues a series, *Digest of statistics,* which includes annual issues on financial data, fleet and personnel, civil aircraft on register, traffic, aircraft accidents, airport traffic, and traffic flow.

ROADS

Road statistics are less comprehensive and less detailed than railway statistics, for instance. They have often been collected for administration controls, such as licensing and taxation, and nearly all refer to the numbers and types of vehicles in use rather than to the amount of traffic on the roads and the volume and type of freight carried. More recently, however, steps have been taken to rectify this imbalance of the information available.

The British Road Federation has published its annual *Basic road statistics* (BRF) since 1934 ' in an attempt to bring some order into the confusion which arises from the use of contradictory or conflicting road statistics '. It brings together the various important types of road statistics, showing runs of annual figures from 1900 or so, but it includes only the most important and in summary form, so that it is very useful for the more general enquiries only.

An important annual statistical publication no longer published was *Tramway and light railway returns* (HMSO, 1877-1938), issued first by the Board of Trade and later by the Ministry of Transport. After the war these returns were continued in *Public road passenger statistics* and later in *Passenger transport in Great Britain,* both of which are referred to earlier in this chapter. The original title was concerned chiefly with trams and trolleybuses, but currently all types of transport are included.

The ministry's annual *Highway statistics* (HMSO, 1963-), which replaced *Road motor vehicles* (HMSO, 1951-62), contains the main statistics relating to the highways of Great Britain and the vehicles that use them, including traffic, mileage and expenditure, transport, vehicles currently licensed and new registrations. *Roads in England* (HMSO, 1956-), until recently titled *Roads in England and Wales* and covering Wales as well as England, is the annual report of the ministry on roads and contains information on road mileages and general capital and revenue expenditure. Similar information for Wales is now given in the annual report of the Welsh Office and for Scotland in the annual report of the Scottish Development Department. Highway expenditure in England and Wales is also in *Highways expenditure of county councils,* issued annually by the Society of County Treasurers and the County Surveyors' Society. Financial information concerning the various municipal passenger transport undertakings is published annually in *Municipal passenger transport, summary of accounts,* by the Municipal Passenger Transport Association, which also includes statistical data on miles operated and passengers carried. Similar, less detailed information appears in the annual review of the periodical *Bus and coach.*

The annual reports of the Road Research Laboratory, *Road research* (HMSO, 1935-) are a source of statistics on road traffic, as are some of their *ad hoc* publications such as *Sample survey of the road and traffic of Great Britain*, by J C Tanner, H D Johnson and J R Scott (HMSO, 1962), which describes the results of a sample enquiry designed to obtain information on road and traffic conditions, and *Research on road traffic* (HMSO, 1965), both of which have tables of road traffic statistics. Sample traffic censuses were taken by the Ministry of Transport in 1954 and 1961 and a full one at 6,000 points in 1965. The results of these censuses have now been published by the ministry in *General traffic censuses (trunk roads), August figures, 1954, 1961 and 1965.*

K F Glover and D N Miller analyse the statistics of road goods transport in an article 'The outlines of the road goods transport industry' (*Journal* of the Royal Statistical Society, series A, vol 117, 1954, 297-330). More recently, the Ministry of Transport made a one week survey of road goods transport in 1962, the results of which are being published in several volumes with the overall title *Survey of road goods transport, 1962* (HMSO, 1966-).

Statistics of road accidents are given in an annual publication, *Road accidents* (HMSO, 1951-), issued jointly by the Ministry of Transport, the Welsh Office and the Scottish Development Department. The Royal Society for the Prevention of Accidents also publishes an annual *Road accident statistics* (ROSPA, 1956-), which contains similar information to the official publication but also information by each police district, and a monthly *Statistical review* containing data on road casualties supplied by the ministry.

The licensing of vehicles is dealt with in the *Annual reports of the licensing authorities to the Ministry of Transport* (HMSO, 1933/34-1937/38; 1947/48-) and the *Annual reports of the traffic commissioners* (HMSO, 1931-), both of which contain statistics showing the number of applications for each type of licence in each area, vehicle examinations, summonses, etc. The Society of Motor Manufacturers & Traders provides the main basic statistical information for the country on the use of motor vehicles. Both the *Monthly statistical review* and the annual *Motor industry of Great Britain* include statistics of motor vehicles in use, classed by type of vehicle and area, new registrations and current licences. These two publications are described in more detail in chapter six.

Statistics of roads, road expenditure and road accidents in Northern Ireland are given in the *Ulster yearbook.*

The main general publication on US road statistics is *Highway statistics* (GPO, 1956-) issued annually by the Department of Commerce's Bureau of Public Roads and including data on motor fuels, motor vehicles, highway user taxation, state and local highway finance, highway mileage, and federal aid for highways. The bureau also issues two quarterly financial publications, *Revenue, expenses, other income, and statistics of Class I motor carriers of passengers* and a similar publication concerned with large motor carriers of property. Non governmental publications containing a wide variety of statistical data are *Automobile facts and figures* and *Motor truck facts,* both issued annually by the Automobile Manufacturers Association, and *Bus facts—a summary of facts and figures on the motor bus industry,* issued annually by the National Association of Motor Bus Owners.

Internationally, the International Road Federation issues an annual *Statistical data on road activity* (IRF, 1965-) and McGraw-Hill's *World automotive market survey* contains data on vehicle registrations for some countries, although mainly concerned with US statistics. The 1966 issue also included a world motor census report. *Statistics of road traffic accidents in Europe* (UN, 1962-) is issued annually by the United Nations Economic Commission for Europe.

COMMUNICATIONS

Statistical tables of letter and parcel post, telegraph and telephone cables, telegrams, telex stations, telephone calls, telephone stations and exchanges are published in the *Annual abstract of statistics.* Numbers of radio receiving and television licences issued annually are also published in the abstract and monthly figures are in the *Monthly digest of statistics.* Some statistics, mainly financial, are included in the reports of the organisations concerned, *ie Report and accounts of the Post Office, Post Office prospects, Report and accounts of Cable and Wireless Ltd, Annual report and accounts of the British Broadcasting Company,* and *Annual report and accounts of the Independent Television Authority,* all published by HMSO.

Much more statistical information has been published in the past and Arthur Hazlewood's article ' Telecommunications statistics ', in *Sources and nature of the statistics of the United Kingdom,* edited by M G Kendall (Oliver & Boyd, 1952-57) gives a very good historical picture of the information that used to be available. A considerable amount of statistical information on telecommunications is collected by the Post Office and the Telecommunications Headquarters, Management Service Department, Post Office Headquarters, St Martin's-le-Grand, London, EC1, can often provide specific unpublished information to enquirers.

In the United States telephone and telegraph statistics are published in the Federal Communications Commission's *Statistics of communications common carriers, Statistics of telephone carriers, reporting annually to Commission* and *Statistics of principal domestic overseas telegraph carriers reporting annually to Commission.* The United States Independent Telephone Association issues an annual *Statistics of the independent telephone industry* and all types of information are available from the American Telephone and Telegraph Company in New York.

UNESCO publishes a series of ' Statistical reports and studies ' which includes some issues on communications, such as number four, *Statistics of newsapers and other periodicals,* and number six, *Statistics on radio and television, 1950-1960* but these are now somewhat out of date. A more recent publication of UNESCO's is *World communications—press, film, radio, television* (UNESCO, fourth edition, 1964). Some information is also included in UNESCO's *Statistical yearbook.* The American Telephone and Telegraph Company in New York publish an annual statistical compilation which is now titled *The world's telephones* (1912-).

CHAPTER ELEVEN

Tourism

The *Annual report of the British Travel Association* (BTA, 1928/29-) contains a few statistics other than the accounts of the association, including the number of holiday visitors and business visitors, by sea and by air, by nationality, and also the total expenditure in this country of each nationality. The Central Statistical Office's *Annual abstract of statistics* has a table showing the number of visitors arriving in the United Kingdom by sea and by air, analysed by nationality, but does not indicate the purpose of the visits (*ie,* for business or pleasure). Brief figures of United Kingdom tourist traffic are given monthly in the *Board of Trade journal,* which also has an annual table on international tourism. The *Annual abstract of statistics* also includes an item on consumers' expenditure abroad in a table showing consumers' expenditure generally, but United Kingdom non merchandise trade is not generally well documented although tourism is shown in the balance of payments statistics. In fact, there is in the United Kingdom nothing to compare with the detailed annual, and often monthly also, volumes of statistical data on tourism published in many Western European countries.

Statistics of tourism in Northern Ireland are given in the *Annual report of the Northern Ireland Tourist Board* (HMSO, Belfast, 1947/48-).

The US Department of Commerce issued a *Travel survey* in 1963 and its United States Travel Service issues six monthly and annually *Pleasure and business visitors to the US by port of entry and mode of travel* in the form of a press handout. And the *Annual report of the United States Immigration and Naturalization Service* (GPO, 1933-) contains statistical tables on immigration, aliens and citizens admitted and departed, and persons naturalised; and the same department issues a semi annual *Report of passenger travel between United States and foreign countries.*

The International Union of Official Travel Organisations, at Geneva, publishes *International travel statistics* (IUOTO, 1947-) annually in which statistical data is given of tourist arrivals and nights spent in each country, the information being analysed by months and by nationality. The Organization for Economic Co-operation and Development issues *Tourism in* OECD *member countries* (OECD, 1962-) annually and it is concerned with the development of international tourism in those countries. Mainly statistical, the figures have been supplied by the various national government offices responsible for the promotion of tourism.

Statistics of tourism are also included in some other publications referred to in the section on migration statistics in chapter two.

Index